Communications in Asteroseismology

Volume 156
November/December, 2008

Austrian Academy
of Sciences Press

Vienna 2008

OAW

Communications in Asteroseismology

Editor-in-Chief: **Michel Breger**, michel.breger@univie.ac.at
Editorial Assistant: **Daniela Klotz**, klotz@astro.univie.ac.at
Layout & Production Manager: **Paul Beck**, paul.beck@univie.ac.at
Language Editor: **Natalie Sas**, natalie.sas@ster.kuleuven.be

Institut für Astronomie der Universität Wien
Türkenschanzstraße 17, A - 1180 Wien, Austria
http://www.univie.ac.at/tops/CoAst/
Comm.Astro@univie.ac.at

Cover Illustration

The Milky Way behind the dome of the 40-inch telescope at the *Siding Spring Observatory, Australia*. Data from this telescope is used in a paper by Handler & Shobbrook in this issue (see page 18).
(Photo kindly provided by R. R. Shobbrook)

All rights reserved
ISBN 978-3-7001-6539-2
ISSN 1021-2043
Copyright © 2008 by
Austrian Academy of Sciences
Vienna

Austrian Academy of Sciences Press
A-1011 Wien, Postfach 471, Postgasse 7/4
Tel. +43-1-515 81/DW 3402-3406, +43-1-512 9050
Fax +43-1-515 81/DW 3400
http://verlag.oeaw.ac.at, e-mail: verlag@oeaw.ac.at

Comm. in Asteroseismology
Vol. 156, 2008

Introductory Remarks

The summer of 2008 was a densely packed season with a number of excellent astero-seismological conferences. CoAst will publish the proceedings of the Wroclaw, Liege and Vienna (JENAM) meetings. In fact, the proceedings of the HELAS Wroclaw conference is mailed to you together with this regular issue. Accompanying these two issues is a user manual for the FAMIAS and DAS packages. The scientific mix of these three simultaneous issues represents the typical CoAst mixture. This also demonstrates the growth of CoAst. Besides this increase in special issues, the number of subscribers of the printed regular issue (free of charge for asteroseismologists presently publishing in the field) has exceeded the magical number of 500. I am especially pleased about our rapid expansion into the Americas.

These days, it is fashionable to judge scientific excellence through numbers. To some extent this is a futile undertaking: how do you judge quality? For scientific papers, the number of citations and the impact factor of the journal are important. Here too, CoAst evolved in a nice way showing a significant increase of the citations per paper. Which papers have been cited the most in previous CoAst issues? Not surprisingly, these were tools of our trade: the PERIOD04 and PERIOD98 handbooks. Therefore, we expect the manual for FAMIAS/DAS (Vol.155) to become highly quoted as well. But not only user manuals for software tools are cited heavily. Our papers on satellite missions have also been heavily read. In the present volume, you will find two papers on pioneering satellite observations. Nevertheless, ground-based astero-seismology is still well-represented.

The CoAst team is also growing and we welcome our latest member: Natalie Sas, who is our language editor supported by *HELAS*. Also, I would like to take this opportunity to thank our contact persons at *ADS*, Carolyn Stern-Grant and Edwin Henneken as well as at *Simbad Database*, Francois Ochsenbein for the excellent collaboration and their support.

Michel Breger
Editor-in-Chief

Contents

Introductory Remarks
by Michel Breger, Editor   3

Scientific Papers

Uncertainties in phase shifts and amplitude ratios: Theory and practice
by M. Breger   6

A new photometric study of the high galactic latitude β Cep star HN Aqr
by G. Handler, and R. R. Shobbrook   13

A three-site photometric campaign on the ZZ Ceti star WD 1524-0030
by G. Handler, J. L. Provencal, M. Lendl, M. H. Montgomery, and P. G. Beck   18

On the nature of HD 207331: a new δ Scuti variable
by L. Fox Machado, W. J. Schuster, C. Zurita, J. L. Ochoa, and J. S. Silva   27

Modeling the Pulsating sdB Star PG 1605+072
by L. van Spaandonk, G. Fontaine, P. Brassard, and C. Aerts   35

Data Reduction pipeline for MOST Guide Stars
and Application to two Observing Runs
by M. Hareter, P. Reegen, R. Kuschnig, W. W. Weiss, J. M. Matthews,
S. M. Rucinski, D. B. Guenther, A. F. J. Moffat, D. Sasselov, and G. A. H. Walker   48

First asteroseismic results from CoRoT
by E. Michel, A. Baglin, W.W. Weiss, M. Auvergne, C. Catala, C. Aerts,
T. Appourchaux, C. Barban, F. Baudin, M. Briquet, F. Carrier, et al.   73

HELAS News

Announcement of HELAS III
by C. Aerts   90

Conference Review of the 38th LIAC / HELAS-ESTA / BAG
Liége, Belgium, July 7-11 2008
by A. Grötsch-Noels, J. Montalban, and A. Miglio   91

HELAS Local Helioseismology Activities
by H. Schunker, and L. Gizon   93

Scientific
Papers

Comm. in Asteroseismology
Vol. 156, 2008

Uncertainties in phase shifts and amplitude ratios: Theory and practice

M. Breger

Astronomisches Institut der Universität Wien,
Türkenschanzstr. 17, A–1180 Wien, Austria

Abstract

Multicolor data of pulsating variables yield information on the amplitude ratios and phase shifts between the different passbands. This is a powerful tool for mode identifications of radial and nonradial pulsators. The identifications rely on relatively small effects, so that the uncertainties due to measurement errors need to be known precisely.

We present the formulae which allow the calculation of the statistical uncertainties from the residuals between the measurements and the least-squares fit of sinusoids to the light curves. Since it has been often presumed that the real uncertainties of the amplitudes and phases are greater than calculated from statistics, we have compared the theoretical scatter with the observed scatter. Nine pulsation modes of the δ Scuti variable 44 Tau, extensively observed for five observing seasons, were chosen. The observed and predicted scatter of 86 pairs of amplitude ratios and phase shifts were compared.

We find that the observed and predicted scatter are very similar: the histograms of the observed scatter in the amplitude ratios and phase shifts match the normal distribution predicted from the formulae. The excellent agreement might be a consequence of the fact that most systematic observational and computational (caused by multiperiodicity) errors tend to cancel out when the measurements at the different passbands are compared.

Individual Objects: 44 Tau

Introduction

The identification of the excited pulsation modes forms the basis of observational asteroseismology. The light curves of the multiperiodic pulsators are generally used to detect the multiple frequencies of pulsation together with their amplitudes and phases. If multicolor light curves are available, these also form powerful tools to identify the pulsation modes (particularly the ℓ values). For the mode identification, two parameters for each frequency are especially important: the amplitude ratios and phase shifts between two carefully chosen passbands. The identifications rely on relatively small effects, so that the uncertainties due to measurement errors need to be known precisely. In fact, only the recent developments of large observational campaigns with high precision have allowed the reliable determination of the small phase shifts.

The solutions to the observed light curves involve multiple sinusoidal fits, which are usually obtained by least-squares algorithms. A number of different statistical packages are used by different research groups. An example is the package PERIOD04 (Lenz & Breger 2005), which computes amplitudes and phases together with the formal uncertainties of these fits. The question arises whether the computed values of the uncertainties are realistic. Montgomery & O'Donoghue (1999) wrote, "the naive least-squares formulae provide only a lower limit to the errors, and the true errors may be much higher." It has been the author's experience that the photometric data obtained for the same stars in different years often show uncertainties up to 50% higher than predicted by the simple formulae (given in the next section). This is due to the fact that observational noise is correlated. In the case of δ Scuti stars, another effect comes into play: the star itself changes the frequency values (i.e., the phases) and amplitudes, often by small amounts. For many years we have been following the star 4 CVn for more than 100 nights/year in order to examine the nature of this variability. The currently unpublished results indicate intrinsic 'jitter' even in the modes with high amplitudes.

When we turn to multicolor data and consider the derived values of amplitude ratios and phase shifts between different passbands, the situation changes. Most two-color measurements are obtained with the same instrument and almost simultaneously. This means that several components of the observational noise cancel out (or their effects are reduced) since they affect both passbands in a similar manner. Of course, photon statistics is not cancelled; this noise source is not correlated from measurement to measurement. Also, the phase and amplitude drifts intrinsic to the star as well as any effect of unrecognized additional frequencies should cancel out. Consequently, the bad situation described in the previous paragraph may not apply to the amplitude ratios and phase differences determined under near-identical conditions for the two colors.

In this paper, we want to compare the predicted and observed scatter in the amplitude ratios and phase shifts for the star 44 Tau: this is one of the few stars (if not the only one) for which five extremely extensive sets of data from different years are available (Breger et al. 2008) and for which the comparison can be carried out.

Theoretical statistical uncertainties

The light curves of many types of pulsating stars (such as δ Scuti stars) are nearly sinusoidal so that they can be mathematically described by a sinusoid with frequency ω_1. The slight asymmetries, in practice, are taken care of by including a $2\omega_1$ term for the modes with the largest amplitudes.

Suppose we have N measurements of the magnitudes, m_i, at times t_i. We assume that the times of the observations are error free, but that the brightness measurements are subject to random errors, Δm_i, which have an average of zero, a constant root-mean-square amplitude, and are not correlated in time.

In order to analyze our time series data, we fit a sinusoid to it. Specifically, we fit the function

$$f(t) = a_0 + a\sin(\omega t_i + \phi),\tag{1}$$

where the frequency, ω, is assumed to be known, but where the amplitude a and phase ϕ need to be determined. This is a realistic situation when we wish to compare the amplitudes and phases for data measured in two passbands. The parameter a_0 represents a constant offset.

We define

$$\begin{aligned}\chi^2 &\equiv \sum_{i=1}^{N}[m_i - f(t_i)]^2\\ &= \sum_{i=1}^{N}[m_i - a_0 - a\sin(\omega t_i + \phi)]^2,\end{aligned}\tag{2}$$

where the minimum in χ^2 corresponds to the best fit solution of the model parameters. We now minimize χ^2 with respect to a_0, a, and ϕ and derive the following relations (for details see Breger et al. 1999, Montgomery & O'Donoghue 1999):

$$\sigma(a) = \sqrt{\frac{2}{N}} \cdot \sigma(m),\tag{3}$$

which is the desired relation between photometric and amplitude uncertainties. Here $\sigma(m)$ is the standard deviation of the measured brightnesses relative to the fits.

We also find

$$\sigma(\phi) = \frac{\sigma(a)}{a} = \sqrt{\frac{2}{N}}\frac{\sigma(m)}{a},\tag{4}$$

which is the desired relation between the photometric error, the amplitude of the signal, and the error in the phase determination.

It is common among observers to express ϕ in degrees and to relate the uncertainties in amplitude and phase. The equation then becomes

$$\sigma(\phi) = 57.3 \; \sigma(a)/a \tag{5}$$

Suppose that the data have been obtained through two separate filters, 1 and 2. Let us assume that the measurements in the two passbands are independent of each other and the errors are not correlated.

When we carry out the error propagation, we find

$$\sigma\left(\frac{a_1}{a_2}\right) = \frac{a_1}{a_2}\sqrt{\left(\frac{\sigma(a_1)}{a_1}\right)^2 + \left(\frac{\sigma(a_2)}{a_2}\right)^2} \tag{6}$$

$$\sigma(\phi_1 - \phi_2) = \sqrt{\left(\frac{\sigma(a_1)}{a_1}\right)^2 + \left(\frac{\sigma(a_2)}{a_2}\right)^2}, \tag{7}$$

or in degrees,

$$\sigma(\phi_1 - \phi_2) = 57.3\sqrt{\left(\frac{\sigma(a_1)}{a_1}\right)^2 + \left(\frac{\sigma(a_2)}{a_2}\right)^2}. \tag{8}$$

This allows us to calculate the uncertainties in the amplitude ratio and phase shift.

Many pulsating stars are multiperiodic so that Eqn. 1 needs to be replaced by

$$f(t) = a_0 + \sum_{j=1}^{F} a_j \sin(\omega_j t_i + \phi_j), \tag{9}$$

where F represents the number of frequencies (and harmonics).

We can only apply the equations derived earlier to each of the frequencies if the data are sufficient to ensure that the different frequencies do not influence each other's solution. This is not fulfilled for small data sets. However, such small data sets would not be used anyhow to give astrophysically meaningful values of phase shifts and amplitude ratios. Montgomery & O'Donoghue (1999) applied a test to six-frequency solutions of the 1996 data of 4 CVn and found a deviation of the errors of no more than 12%. Since most of the recent available campaigns are much more extensive than the data set used by Montgomery & O'Donoghue, we do not consider the multiple frequencies as a severe problem.

Table 1: Example of a comparison of phase shifts (radial mode at 8.96 c/d)

Observing season	Phase shift degrees	Difference, d, between years degrees	normalized
2004/5	2.93 ± 1.02		
2005/6	2.46 ± 0.93	0.47 ± 1.38	0.34

Observed uncertainties in in phase shifts and amplitude ratios

Recently, a very large data set has become available, which allows us to examine the validity of the uncertainties computed from the residuals between the measurements and the fits, i.e., δm. The star 44 Tau was studied extensively for five years from 2000/1 to 2005/6 (Breger & Lenz 2008) in two colors, viz., the Stromgren v and y passbands. We selected nine frequencies with relatively large amplitudes. A tenth frequency at 9.56 c/d was not used because of the presence of a close frequency companion at 9.58 c/d and the possible contamination. Some of the frequencies show strong amplitude variability from year to year. This caused us to reject one of the forty-five available solutions: in 2005/6, the mode at 9.12 c/d had a near-zero amplitude.

For each observing season and frequency, we computed the amplitude ratio, v/y, and phase shift, $\phi(v) - \phi(y)$. We also calculated the formal uncertainties of these values, which differ because of the different number of observations, annual residuals of the fits and variable amplitudes of pulsation. We then compared the results for each year with those of each of the other years. This resulted in ten comparisons for each frequency. The differences in the measured values, d, were then normalized to the 'expected' standard deviations computed from the known residuals of the annual solutions, $\sigma(m)$. We obtained 86 ratios for each of the amplitude ratios and phase shifts. These should ideally follow a normal distribution. Our approach is illustrated in Table 1, which shows one of the 86 calculations.

We then examined the distribution of the 86 amplitude-ratio and phase-shift differences. Since each value was already scaled (normalized) to the predicted standard deviation, a perfect fit of the statistical formulae discussed in the previous section would lead to a normal distribution of these 86 values. The actual distributions are shown in Fig. 1 and 2. The bin size was chosen to give a sufficiently high number of occurrences, B, in order for the uncertainties $(=\sqrt{B})$ not to dominate. We found that, visually, the agreement with the predicted normal distribution (shown as curves) is excellent. If we fit Gaussian curves to the data and examine the standard deviations corresponding to these

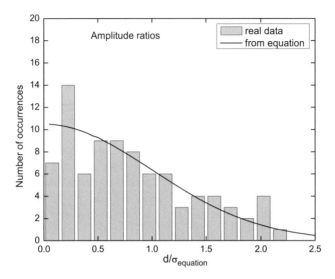

Figure 1: Histogram of the observed differences in the calculated amplitude ratios between different years for nine frequencies of 44 Tau. The differences were normalized relative to the standard deviations expected from the theoretical uncertainties. The drawn curve represents the expected Gaussian distribution. Each observed number of occurrences has a formal uncertainty of the square root of its value.

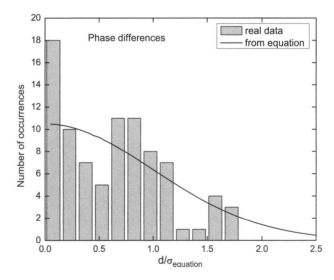

Figure 2: Same as Fig. 1 for the observed phase shift, $\phi(v) - \phi(y)$. This diagram shows that the observed and theoreticallypredicted scatter are similar.

curves, the amplitude ratios have slightly larger standard deviations, while the phase shifts show slightly smaller values. In both cases, the deviations from 1.0 are not statistically significant.

Finally, it needs to be noted that the present examination dealt with the statistics of phase shifts and amplitude ratios, when the variations in different passbands are compared. Since these two-color measurements are collected almost simultaneously with the same instrument, many systematic observational errors and computational errors (caused by multiperiodicity) affect both sets of measurements similarly. They, therefore, tend to cancel out. This is not the case for the light curves in a single color: the absolute values of the amplitudes and phases of the fit may still be less accurate than predicted by the formulae.

We conclude that for the data sets tested, the theoretical uncertainties of the amplitude ratios and phase shifts correspond very well to the observed values. When the measurements in the two colors are obtained under similar observing conditions, there is no need to artificially increase the values of the computed statistical uncertainties.

Acknowledgments.   It is a pleasure to thank Michael Endl for interesting discussions. This investigation has been supported by the Austrian Fonds zur Förderung der wissenschaftlichen Forschung.

References

Breger, M., & Lenz, P. 2008, A&A, 488, 643

Breger, M., Handler, G., Garrido, R., et al. 1999, A&A, 349, 225

Lenz, P., & Breger, M. 2005, CoAst, 146, 53

Montgomery, M. H., & O'Donoghue, D. 1999, DSSN, 13, 28

Comm. in Asteroseismology
Vol. 156, 2008

A new photometric study of the
high galactic latitude β Cep star HN Aqr

G. Handler[1], and R. R. Shobbrook[2]

[1] Institut für Astronomie, Türkenschanzstrasse 17, A-1180 Vienna, Austria
[2] Research School of Astronomy and Astrophysics, Australian National University,
Canberra, ACT, Australia

Abstract

We have carried out 32 h of new time-resolved CCD UBV photometry of the high galactic latitude β Cephei star HN Aqr. We detected its known single pulsation frequency and noticed additional variability at lower frequencies. It is not clear whether or not this longer-term variation originates from HN Aqr itself. The UBV colour amplitudes of the star's β Cephei pulsation point towards an $\ell = 4$ or $\ell = 2$ mode, inconsistent with previous mode identifications.

Individual Objects: HN Aqr

Introduction

The β Cephei stars are a group of early B-type stars with masses between 9 and 17 $M_\odot$ that exhibit light, radial velocity and line-profile variability on time scales from two to eight hours (Stankov & Handler 2005). Their variability is caused by pulsations in pressure and gravity modes of low radial order that are driven in the ionization zone of the iron-group elements (Moskalik & Dziembowski 1992). As a direct consequence, the strength of the pulsational driving of these variables should be directly related to the abundance of the iron-group elements in the driving zone, and the pulsations should vanish if the metallicity (for a given element mixture) falls below a certain limit (e.g. Pamyatnykh 1999).

Nevertheless, numerous β Cephei stars were reported in the generally metal-poor Large Magellanic Cloud (Kołaczkowski et al. 2004). Still, this finding may be explicable by the presence of regions with higher metallicity within the LMC. Additional relief from the theoretical side may be supplied by the recent revision of the solar element mixture (Asplund et al. 2005) causing a *relative* increase of the abundances of the iron-group elements with respect to CNO that dominates determinations of the overall metallicity of early-type stars (Pamyatnykh 2007).

There is a single case of a galactic β Cephei star in a low-metallicity environment: HN Aquarii (=PHL 346). Discovered to pulsate by Waelkens & Rufener (1988) and confirmed by Kilkenny & Van Wyk (1990), this object has long been taken as evidence for star formation in the galactic halo (e.g. Keenan et al. 1986). However, proper motion measurements made an explanation in terms of a runaway star from the galactic plane possible (Ramspeck et al. 2001).

Asteroseismology may shed additional light on this problem. Pamyatnykh et al. (2004) showed that some pulsation modes are sensitive to the effects of metallicity on interior stellar structure and can therefore be used as a seismic probe of the overall interior metal abundance.

So far, only a single mode of pulsation has been detected for HN Aqr, but the published data sets are rather small by today's standards. The noise level in the prewhitened data of Kilkenny & van Wyk (1990), the most extensive photometric study of HN Aqr published to date, is 3.5 mmag. Heynderickx et al. (1994) reported a mode identification for the strongest pulsational signal of HN Aqr: by combining ultraviolet, Walraven and Geneva photometry, they suggested a spherical degree of $\ell = 1$ as the most likely. To improve the observational data base on HN Aqr, we decided to carry out new measurements in order to detect previously unknown pulsation modes and to provide mode identifications.

Observations and reductions

We used the 1.0-m telescope at the Siding Spring Observatory and the Wide-Field Imager (WFI) to acquire time-resolved UBV CCD photometry of HN Aqr. We measured the star in July and September 2006 during six nights each, totalling 32.2 hours of observation.

The data were reduced with standard IRAF[1] routines, and were corrected for overscan, flat field and nonlinear response of the CCD chip (only WFI chip #3, which had the best quality, was used); bias and dark count corrections were not found necessary. Photometry was carried out with the program package MOMF (Multi–Object Multi–Frame, Kjeldsen & Frandsen 1992) that applies combined Point–Spread Function/Aperture photometry relative to an optimal sample of comparison stars, ensuring highest-quality differential target light curves.

---

[1] IRAF, the Image Reduction and Analysis Facility, is written and supported by the IRAF programming group at the National Optical Astronomy Observatories (NOAO) in Tucson, Arizona.

Light curve and frequency analysis

We searched the data for periodicities using the program Period04 (Lenz & Breger 2005). This package applies single-frequency power spectrum analysis and simultaneous multi-frequency sine-wave fitting. The amplitude spectrum of our V filter data is shown in Fig. 1.

This amplitude spectrum is dominated by two features: the known pulsational signal near 6.5 c/d plus some low-frequency variability. The latter dominates the residual amplitude spectrum after prewhitening the known pulsation frequency, leading to the disappointing result that no further β Cephei-type oscillations can be revealed in our data. Adopting the period derived by Kilkenny & van Wyk (1990) to resolve aliasing ambiguities, but optimizing it to accommodate possible frequency variability that may have occurred in the 18 years between the two data sets, results in a pulsation frequency of 6.5666 ± 0.0004 c/d in our data[2]. The high formal accuracy quoted is a consequence of the two-month time baseline of our data. The UBV amplitudes of this pulsation mode are 21.6 ± 2.0, 21.2 ± 1.1, and 20.8 ± 1.0 mmag, respectively; the phase of the pulsational signal is the same at all wavelengths within the errors.

With these amplitudes, the spherical degree ℓ of the pulsation mode can be constrained. A comparison between the observed and theoretically predicted UBV amplitudes for HN Aqr is shown in Fig. 2. We note that the measured amplitudes are most consistent with the prediction for an $\ell = 4$ mode, although $\ell = 2$ cannot be ruled out keeping in mind the possibility of systematic errors. Other ℓ values are difficult to be reconciled with our data.

Turning to the low-frequency variability, we cannot clearly say whether or not it is intrinsic to HN Aqr. Analysing the differential photometry of other stars of similar brightness in the field in the same way did not result in such low-frequency variability, suggesting that HN Aqr is responsible for it. On the other hand, the differential magnitudes of HN Aqr are correlated with its x-position on the chip (and are not correlated with y-position or seeing). A corresponding decorrelation did not fully remove the low-frequency signal.

Discussion

Our new CCD observations of the high-galactic latitude β Cephei pulsator HN Aqr did not suffice to deepen our understanding of the star. However, two questions turned up that may make a more in-depth study worthwhile.

First, it is unclear where the low-frequency variations in our light curves originate from. Interestingly, Kilkenny & van Wyk (1990) discussed a similar

---

[2]Kilkenny (private communication) reports a period of 0.1523135 d from the published data supplemented by new observations from the following season.

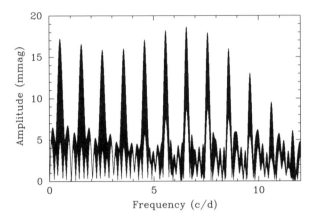

Figure 1: Amplitude spectrum of our V filter photometry of HN Aqr.

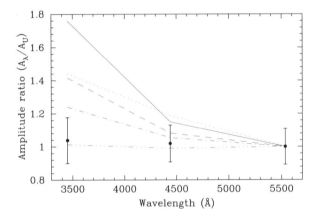

Figure 2: Mode identifications for HN Aqr from a comparison of observed and theoretical UBV amplitude ratios, normalized to unity at V (where the relative error of amplitude determination is smallest). The filled circles with error bars are the observed amplitude ratios. The full lines are theoretical predictions for radial modes, the dashed lines for dipole modes, the dashed-dotted lines for quadrupole modes, the dotted lines for octupole modes, and the dashed-dot-dot-dotted lines are for $\ell = 4$.

problem in their observations of HN Aqr, but they suspected that their single comparison star could be responsible for the slow variability. On the other hand, Waelkens & Rufener (1988) as well as Heynderickx (1992) did not mention low-frequency variations, although we have to mention that their data sets were less extensive than ours and Kilkenny & van Wyk's (1990).

Second, our mode identification does not agree with that by Heynderickx et al. (1994). Again, it is hard to pinpoint the cause of this disagreement. The previously mentioned authors did not give error estimates for the photometric amplitudes used. Perhaps the low-frequency variability went unnoticed in the small data sets available to them, but did affect the amplitude determinations systematically. The same comment may be applicable to our data.

One might also suspect that our U passband does not conform to the standard system. Therefore, we compared the UBV amplitudes of some β Cephei stars with those obtained from other sites and found them to be consistent within the errors. Nevertheless, we point out the necessity for well-defined U filter passbands when attempting to identify modes of hot stars from CCD photometry; the ultraviolet amplitudes are crucial for correct identifications.

We leave the reader with the standard trivial conclusion that follows many observational studies: more data are needed.

Acknowledgments. This research was supported by the Austrian Fonds zur Förderung der wissenschaftlichen Forschung under grant P18339-N08.
GH thanks Vichi Antoci, Dave Kilkenny and the referee for helpful discussions.

References

Asplund, M., Grevesse, N., & Sauval, A. J. 2005, ASPC, 336, 25

Heynderickx, D. 1992, A&AS, 96, 207

Heynderickx, D., Waelkens, C., & Smeyers, P. 1994, A&AS, 105, 447

Keenan, F. P., Lennon, D. J., Brown, P. J. F., & Dufton, P. L. 1986, ApJ, 307, 694

Kilkenny, D., & van Wyk, F. 1990, MNRAS 244, 727

Kjeldsen, H., & Frandsen, S. 1992, PASP, 104, 413

Kołaczkowski, Z., Pigulski, A., Soszyński, I., et al. 2004, ASPC, 310, 225

Lenz, P., & Breger, M. 2005, CoAst, 146, 53

Moskalik, P., & Dziembowski, W. A. 1992, A&A, 256, L5

Pamyatnykh, A. A. 1999, AcA, 49, 119

Pamyatnykh, A. A. 2007, CoAst, 150, 207

Pamyatnykh, A. A., Handler, G., & Dziembowski, W. A. 2004, MNRAS, 350, 1022

Ramspeck, M., Heber, U., & Moehler, S. 2001, A&A, 378, 907

Stankov, A., & Handler, G. 2005, ApJS, 158, 193

Waelkens, C., & Rufener, F. 1988, A&A, 201, L5

Comm. in Asteroseismology
Vol. 156, 2008

A three-site photometric campaign on the ZZ Ceti star WD 1524-0030

G. Handler,[1] J. L. Provencal,[2,3] M. Lendl,[1]
M. H. Montgomery,[3,4] and P. G. Beck[1]

[1] Institut für Astronomie, Türkenschanzstrasse 17, A-1180 Vienna, Austria
[2] Dept. of Physics and Astronomy, University of Delaware,
223 Sharp Laboratory, Newark, DE 19716, USA
[3] Delaware Asteroseismic Research Center,
Mt. Cuba Observatory,Greenville, DE 19807, USA
[4] The University of Texas at Austin, Department of Astronomy,
1 University Station C1400, Austin, TX 78712, USA

Abstract

We obtained 74 hours of time-resolved CCD photometry of the pulsating DA white dwarf star WD 1524-0030 from three different sites well separated in longitude. We found evidence for amplitude variability with relative changes of $\pm 10\%$ and detected a total of 15 independent and 10 combination frequencies in our light curves. The large number of excited modes, the high amplitudes and nonsinusoidal light curves, the apparent brightness and the equatorial location on the sky make WD 1524-0030 an attractive target for future campaigns with the goal of asteroseismology and nonlinear light curve fitting.

Individual Objects: WD 1524-0030

Introduction

Asteroseismology of pulsating DA white dwarf stars (the ZZ Ceti stars), in the sense of exploring their interior structures in very detail, has always been difficult. The sparse and temporally unstable mode spectra, combined with an interior structure more complicated than that of the DB and PG 1159 pulsators, have been a major obstacle to the study of cool white dwarf interiors.

The strategy that is generally applied to overcome these problems is to obtain observational mode identifications. Spectroscopy can lead to such identifications (e.g., see Thompson et al. 2008 and references therein), but requires very large telescopes to reach the necessary signal-to-noise ratio because pulsating white dwarf stars are intrinsically and apparently faint ($V > 12.2$) and oscillate rapidly (100 s $\lesssim P \lesssim$ 1000 s).

However, another method for mode identification may have come to the rescue. Brickhill (1992) showed how the surface convection zone of pulsating white dwarfs modifies their light curve shapes, causing harmonic and combination signals in frequency spectra. Following up on this work, Wu (1998) pointed out that the amplitudes and phases of these combination terms depend on the types of pulsation modes that cause them. Consequently, Montgomery (2005) derived a method to constrain pulsational mode identifications from the light curve shapes of pulsating white dwarfs, and to recover the thermal response time scale of the convection zone, which depends on effective temperature.

This is excellent news for asteroseismology of the cooler DB and DA white dwarf stars: if several pulsation modes are observed in some of these pulsators, and if their light curve shapes are nonsinusoidal and of reasonably large amplitude, mode identifications can be derived and compared with those from other methods, like pattern recognition or spectroscopy.

The Delaware Asteroseismic Research Center (DARC, Provencal & Shipman 2006), home of the Whole Earth Telescope (WET, Nather et al. 1990), has recently concentrated on exploiting this method. In May 2006, the prototype DB pulsator GD 358 was observed (Provencal et al. 2008), and the ZZ Ceti star EC 14012-1446 was the subject of a recent WET run in March/April 2008. Furthermore, additional targets for simultaneous nonlinear light-curve fitting and asteroseismology are searched for via smaller campaigns. Ideally, such an object is bright, has high pulsation amplitudes, nonsinusoidal light curves, and many modes. One of these candidates is the ZZ Ceti star WD 1524-0030, discovered by Mukadam et al. (2004). The present paper reports on the results of an exploratory campaign devoted to WD 1524-0030.

Observations and reductions

In March/April 2007, we obtained time-resolved CCD photometry of WD 1524-0030 with three telescopes at different sites: the 0.9-m telescope at the Cerro Tololo Interamerican Observatory (CTIO) in Chile, the 0.8-m telescope at the Vienna University Observatory in Austria, and the 2.1-m telescope at the McDonald Observatory in Texas, USA. All measurements were acquired through a BG40 or S8612 filter, serving three purposes: first, to have similar wavelength response, close to that of a blue photomultiplier tube; second, to reduce effects

Table 1: Journal of the observations

Observatory	Run start date/time dd/mm hh:mm (UT)	Length hr	#data points
CTIO	30/03 04:27	5.32	646
CTIO	01/04 04:01	6.26	685
CTIO	03/04 03:49	6.31	762
CTIO	06/04 05:06	4.79	493
CTIO	08/04 03:42	5.86	568
CTIO	10/04 06:20	3.67	391
Vienna	11/04 22:18	4.96	843
Vienna	12/04 22:28	4.74	800
Vienna	13/04 22:23	4.42	722
Vienna	14/04 22:21	4.65	791
Vienna	15/04 22:24	4.56	713
McDonald	22/04 10:39	1.32	923
McDonald	24/04 06:19	5.73	2065
McDonald	25/04 05:46	6.06	2169
McDonald	26/04 06:18	5.56	1092
Total		74.21	13663

of differential colour extinction; third, to suppress the contribution of the visual companion star of WD 1524-0030, a very red object, to the light curves. Table 1 contains the journal of the observations.

The data were reduced with standard IRAF procedures to correct the images for overscan, bias level (if needed), dark counts (if needed) and flat field. Photometry of the Vienna data was carried out using the MOMF (Multi–Object Multi–Frame, Kjeldsen & Frandsen 1992) package, whereas the other measurements were subjected to a series of IRAF scripts employing aperture photometry optimized for high-speed CCD data (Kanaan et al. 2002). Both photometry packages give results of comparable quality, in the form of a differential light curve of the target star.

Given the difference in the size of the telescopes used, and given the fact that observatories at mountain sites, but also in an urban area were involved, one might wonder if the quality of the light curves are comparable. Some examples are shown in Fig. 1. Whereas the data from McDonald Observatory are clearly best, the oscillations of this 15th magnitude star are still conspicuously present in the light curves from Vienna. We note that we plotted the fractional intensity variation of the star, in units of modulation intensity, and that we will use the modulation amplitude as our unit of choice for expressing amplitudes (see WET 1993 for definitions) to avoid the logarithmic scale of magnitudes, and the consequent unphysical light curve distortions.

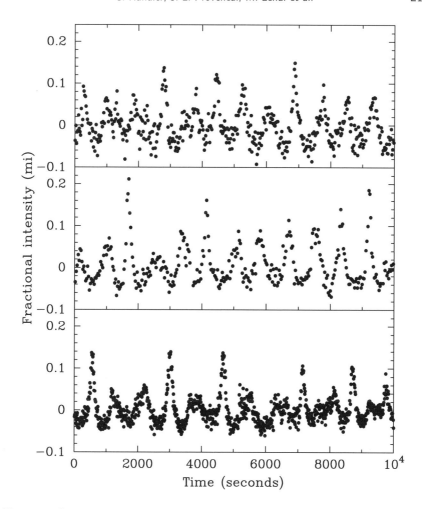

Figure 1: Some example light curves of WD 1524-0030. Upper panel: data from Vienna (April 12), from a 0.8-m telescope at an urban site. Middle panel: a light curve from CTIO (April 6), from a 0.9-m telescope at a mountain site. Lower panel data from McDonald Observatory (April 25), taken with a 2.1-m telescope at a mountain site. Note that the latter data have twice the sampling rate compared to the rest.

For the last reduction step, the light curves were combined. To give all data equal weight in the frequency analysis to follow, the measurements from McDonald Observatory were binned to the same cadence as the CTIO and Vienna data (20 s). Finally, all timings were converted to Barycentric Julian Ephemeris Date (BJED) to ensure a common and consistent time base.

Frequency analysis

We searched the data for periodicities using the program Period04 (Lenz & Breger 2005). This package applies single-frequency power spectrum analysis and simultaneous multi-frequency sine-wave fitting. It also includes advanced options such as the calculation of optimal light-curve fits for multi-periodic signals including harmonic and combination frequencies, which will be required in our analysis.

The amplitude spectrum of our combined light curves is presented in Fig. 2. There are several regions that contain significant power, and the spectral window function is in most cases simpler than the structure in the regions of power, indicating the presence of more than one signal.

Because pulsating DA and DB white dwarf stars in the cooler domains of their instability strip often show amplitude and frequency variations on short time scales, we first computed the Fourier amplitude spectrum of the data from the different sites, which are also separated in time (cf. Table 1). The periodogram from the measurements at the McDonald Observatory showed a maximum amplitude about 20 per cent higher than the periodograms from the other two sites, indicating possible amplitude variability.

We therefore used the following approach to frequency search: we examined three subsets of the data separately, and only accepted signals that were convincingly present in all three. The subsets were chosen with respect to their temporal distribution (see Table 1): a) all data (best time resolution and spectral window), b) the Vienna data plus the last night from CTIO (highest duty cycle), c) all CTIO and Vienna data (good spectral window and duty cycle: the McDonald observations commenced one week after Vienna finished). With this strategy we detected fourteen independent frequencies and the harmonic of the strongest signal.

With these frequencies a rough examination of the presence of amplitude and/or frequency variations during our observations is possible. We assumed constant frequencies over the whole data set and fitted them individually to the data from the three sites, leaving only the amplitudes and phases as free parameters. Whereas we found little evidence for temporal changes in the phases (and thus frequencies), some amplitude changes were detected. Because the amplitudes did not change systematically from site to site, we can rule out that they are affected by possible residual differences between the photometric passbands, and thus different contributions from the visual companion, within the accuracy of our measurements. The low duty cycle of our measurements precludes a deeper investigation of amplitude/frequency variability, but we can surely state that they must be intrinsic to the star, and are in most cases not due to beating between signals unresolved in frequency.

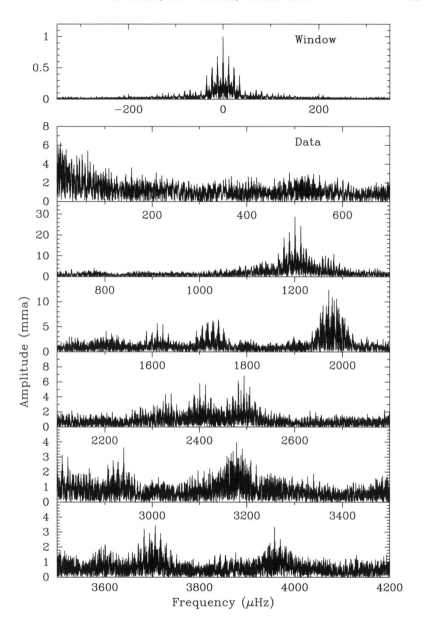

Figure 2: Amplitude spectrum of our multisite data of WD 1524-0030. The uppermost panel is the spectral window function of the data.

Table 2: Frequencies and mean amplitudes of the signals in our light curves of WD 1524-0030, not corrected for the influence of the close companion star. We cannot give reasonable error estimates because of the amplitude/frequency variations during our measurements.

ID	Freq. (μHz)	Ampl. (mma)	Period (s)	ID	Freq. (μHz)	Ampl. (mma)
Independent frequencies				Combination signals		
f_1	1139.31	6.8	877.7	$f_{10} - f_3$	777.80	4.0
f_2	1189.81	9.4	840.5	$2f_3$	2400.72	5.4
f_3	1200.36	25.7	833.1	$f_3 + f_4$	2481.83	4.2
f_4	1281.47	9.4	780.3	$f_3 + f_6$	2811.10	3.1
f_5	1502.54	4.2	665.5	$f_3 + f_9$	3171.07	3.0
f_6	1610.75	5.8	620.8	$f_3 + f_{10}$	3178.51	3.9
f_7	1718.35	5.1	582.0	$f_3 + f_{12}$	3187.73	2.9
f_8	1728.21	5.9	578.6	$2f_3 + f_4$	3682.19	2.1
f_9	1970.71	10.9	507.4	$f_7 + f_{12}$	3705.72	2.5
f_{10}	1978.15	10.1	505.5	$f_9 + f_{12}$	3958.09	3.3
f_{11}	1979.91	5.2	505.1			
f_{12}	1987.38	8.4	503.2			
f_{13}	2340.06	4.4	427.3			
f_{14}	2494.25	5.4	400.9			
f_{15}	2940.09	3.5	340.1			

We continued the analysis by examining amplitude spectra prewhitened by the previously detected signals, keeping the frequencies and phases constant, but letting the amplitude vary for each site. However, keeping in mind the limitations of our data set, we concentrated on finding combination frequencies rather than independent signals. A single convincing independent frequency was added to our result, and another nine combination signals were found. The final list of frequencies obtained in this way is given in Table 2, and the residual amplitude spectrum after their prewhitening is the subject of Fig. 3.

Our frequency solution has succeeded in explaining most of the variability measured, with the notable exception of the frequency region around 1200 μHz, in which the periodogram is quite complicated. The broad (about 200 μHz wide) envelope of this residual structure suggests the presence of several additional pulsational signals: it is wider than anything possibly produced by a combination of a poor spectral window and amplitude/frequency variations on time scales of about one week and it is broad enough to possibly comprise about five radial overtones of $\ell = 1$ modes.

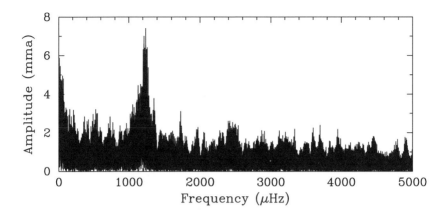

Figure 3: Residual amplitude spectrum of our data of WD 1524-0030 after prewhitening the 25 signals from Table 2.

However, we did not dare to push the frequency analysis further in this domain because we feared running into aliasing problems and into the realm of speculation. In addition, amplitude variations of pulsating white dwarf stars are generally more pronounced at higher radial overtones (see Handler et al. 2008 or Provencal et al. 2008 for discussions), and therefore lower frequencies, and we cannot expect our crude method to take them reliably into account in this case. We also note the occurrence of two peaks (f_{10} and f_{11}) spaced by only 1.8 μHz, but our data set does not allow us to investigate whether these are two independent signals or this might be an artefact caused by amplitude variability.

Discussion and conclusions

We have acquired over 70 h of time-resolved CCD photometry of the pulsating white dwarf star WD 1524-0030 from three sites. We detected fifteen independent signals in its light curves that should be caused by independent pulsation modes. Seven of these cause combination frequencies, making them even more likely to be independent modes. Sometimes we found multiple peaks, notably in the 1980 μHz region, that could be an indication of rotational splitting. One might also speculate about the presence of an ≈ 40 s period spacing among the independent signals.

Our three-site campaign has succeeded to show that WD 1524-0030 is a good target for applying asteroseismic methods to this ZZ Ceti star. In addition, its high amplitudes and nonsinusoidal light curves make it an interesting target to apply Montgomery's (2005) nonlinear light curve fitting method. Such an effort has however to be based on a considerably more extensive data set, which could ideally be provided by the Whole Earth Telescope.

Acknowledgments. This research was supported by the Austrian Fonds zur Förderung der wissenschaftlichen Forschung under grant P18339-N08, Mt. Cuba Observatory and the Delaware Asteroseismic Research Center. GH thanks the referee Barbara Castanheira for her helpful comments.

References

Brickhill, A. J. 1992, MNRAS, 259, 519

Handler, G., Romero-Colmenero, E., Provencal, J. L., et al. 2008, MNRAS, 388, 1444

Kanaan, A., Kepler, S. O., & Winget, D. E. 2002, A&A, 389, 896

Kjeldsen, H., & Frandsen, S. 1992, PASP, 104, 413

Lenz, P., & Breger, M. 2005, CoAst, 146, 53

Montgomery, M. H. 2005, ApJ, 633, 1142

Mukadam, A. S., Mullally, F., Nather, R. E., et al. 2004, ApJ, 607, 982

Nather, R. E., Winget, D. E., Clemens, J. C., et al. 1990, ApJ, 361, 309

Provencal, J. L., & Shipman, H. L. 2006, CoAst, 150, 293

Provencal, J. L., Montgomery, M. H., Kanaan, A., et al. 2008, ApJ, in press

Thompson, S. E., van Kerkwijk, M. H., & Clemens, J. C. 2008, MNRAS, in press

WET 1993, Baltic Astronomy, 2, 568

Wu, Y. 1998, PhD thesis, California Institute of Technology

Comm. in Asteroseismology
Vol. 156, 2008

On the nature of HD 207331: a new δ Scuti variable

L. Fox Machado[1], W. J. Schuster[1], C. Zurita[2], J. L. Ochoa[1], and J. S. Silva[1]

[1] Observatorio Astronómico Nacional, Instituto de Astronomía, Universidad Nacional Autónoma de México, Ensenada B.C., Apdo. Postal 877, Mexico,
[2] Instituto de Astrofísica de Canarias, E-38205 La Laguna, Tenerife, Spain

Abstract

While testing a Strömgren spectrophotometer attached to the 1.5 m telescope at the San Pedro Mártir observatory, Mexico, a number of A-type stars were observed, one of which, HD 207331, presented clear indications of photometric variability. CCD photometric data acquired soon after, confirmed its variability. In order to determine its pulsation behaviour more accurately, $uvby$ differential photoelectric photometry was carried out for three nights. As a result of the period analysis of the light curves we have found a dominant pulsation mode at 21.1 cd^{-1} with an amplitude of 6 mmag. This strongly suggests that HD 207331 is a new δ Scuti-type pulsating star.

Individual Objects: HD 207331, BD +42 4208, TYC 3196-1243-1, HD 208310, HD 209113

Introduction

The δ Scuti-type pulsators are stars with masses between 1.5 and 2.5 $M_\odot$ located at the intersection of the classical Cepheid instability strip with the main sequence. These variables, pulsating with radial and nonradial modes excited by the κ mechanism, are considered to be excellent laboratories for probing the internal structure of intermediate mass stars. Thus, any new detection of a δ Scuti star can be a valuable contribution to asteroseismology.

HD 207331 ($=$ SAO 51294, BD+42 4207, HIP 107557) is classified in the SIMBAD database as a normal A0 star with apparent magnitude V of 8.31 mag. The Hipparcos catalogue (Perryman et al. 1997), on the other hand, lists an H_p of 8.3970 $\pm$ 0.0022 mag (median error), $\pm$ 0.019 mag (scatter), a V_T of 8.335 $\pm$ 0.009 mag (standard error), and a $(B-V)_J$ of 0.217 $\pm$ 0.011 mag

Table 1: Position, magnitude and spectral type of target, comparison and check stars observed in the CCD frame.

Star	ID	RA (2000.0)	Dec (2000.0)	V (mag)	SpTyp
Target	HD 207331	21 47 02	+43 19 19	8.3	$A0$
Comp.	BD+42 4208	21 47 12	+43 19 51	9.4	$A0$
Check	TYC 3196-1243-1	21 47 06	+43 18 58	10.9	-

HD 207331 1:comp. 2:check FOV=5'x5'

Figure 1: The finding chart of HD 207331 (the central star). 1 stands for the comparison star, and 2 for the check star. Some properties of the stars are listed in Table 1. North is up and East is left.

(standard error). Although it may present a possible variability given that $H_{p,\max} = 8.37$ mag and $H_{p,\min} = 8.43$ mag reported in this catalogue, no period nor classification of variability was reported to date. In the present paper, a study of the photometric variability of HD 207331 is reported.

Observations and data reduction

The observations were carried out at the Observatorio Astronómico Nacional, San Pedro Mártir, Baja California, Mexico. The photometric variability of HD 207331 was established on the night of September 27, 2007, using the six-channel $uvby - \beta$ spectrophotometer attached to the H.L. Johnson 1.5 m telescope. The few data clearly show indications of photometric variability for this star.

CCD differential photometry

CCD photometric observations of HD 207331 confirming its variability were carried out on the night of September 30, 2007, with the 0.84 m telescope. A 1024×1024 CCD camera was used with a plate scale of $0.43''$/pixel. About 4 h of data were obtained using a Johnson B filter.

Fig. 1 shows the finding chart of HD 207331. The coordinates, V magnitudes and identifications of the stars marked as 1 and 2 in Fig. 1 are listed in Table 1. The acquired images were reduced in the standard way using the IRAF package. Aperture photometry was applied to extract the instrumental magnitudes of the stars. The differential magnitudes were normalized by subtracting the mean of differential magnitudes for the night.

The differential light curve HD 207331 - Comparison is illustrated in Fig. 2. An exposure time of about 15 s was applied (these data are depicted with dots in Fig. 2). The 3 min binned data are shown by asterisks. As can be seen, the oscillations of HD 207331 are clearly inferred. The magnitude differences between the comparison and check stars were also derived in order to confirm their constancy. No indications of photometric variability for these stars were found.

$uvby$ differential photometry

The star was also observed during three nights in November 2007. The observations were performed with the 1.5 m telescope and the six-channel Strömgren spectrophotometer (Schuster & Nissen 1988). The observing routine consisted of five 10 s integrations of the star from which five times one 10 s integration of sky was subtracted. Two constant comparison stars were observed as well. The star was monitored for about 3.5 h on November 11, for about 4 h on November 18, and 3.5 h on November 19. A set of standard stars was also observed each night to transform instrumental observations onto the standard system. Figure 3 shows the differential light curves in the y filter of HD 207331 for the three nights of our observations. The comparison star C1 is HD 208310 (=BD+44 3980, $V = 8.43$, spectral type A0), while the comparison star C2

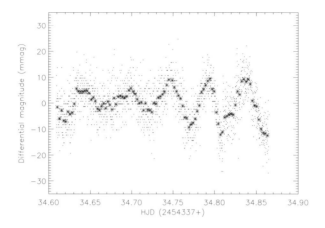

Figure 2: CCD differential light curve HD 207331 - Comparison

corresponds to HD 209113 (=BD+44 4012, $V = 8.42$, spectral type A2). The light curves $HD\,207331 - C1$ and $HD\,207331 - C2$ are shown in the three top and three middle plots of Fig. 3 respectively, with appropriate correction for atmospheric extinction. A multiperiodic characteristic of the star with at least two beating periods can be inferred from these light curves. The CCD differential photometry already showed such behaviour for the light curve of HD 207331 (see Fig. 2).

The magnitude differences between the comparison stars, $C1 - C2$, were also derived in order to confirm their constancy. As can be seen in Figure 3 (the three bottom plots), no indications of photometric variability for these stars were found.

Absolute Strömgren photometry

Preliminary standard photometry of HD 207331, taken on the same nights as the differential photometry, is the following (in mag): $(V, b-y, m_1, c_1) = (8.329, 0.125, 0.150, 1.018)$ with the standard errors of a single observation being $(\pm0.005, 0.002, 0.002, 0.008)$ for 214 observations over the three independent nights.

To deredden this photometry, the β index is not very useful since HD 207331 falls in the spectral range A0-A3, near the maximum of the hydrogen-line absorption, where this index is not very sensitive, changing from merely a temperature indicator for the latter stellar types to merely a luminosity indicator for the

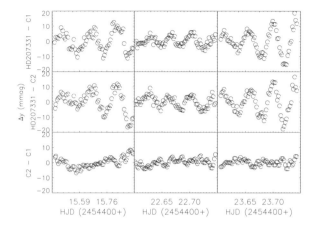

Figure 3: The y differential light curves with respect to the reference stars C1 and C2 (top and middle plots, respectively). The bottom plots are for C1 - C2.

O- and B-type stars (Crawford 1978, 1979). So, a first estimate for the reddening, $E(B - V)(l, b)_\infty$, has been taken from the reddening maps of Schlegel et al. (1998) via the web-page calculator of the NED (NASA/IPAC Extragalactic Database). This reddening is then reduced by a factor $\{1 - exp[-d\sin|b|/H]\}$, where b and d are the Galactic latitude and distance, respectively, assuming that the Galactic dust layer has a scale height $H = 125$ pc. According to SIMBAD $b = -7.823°$ for HD207331, and for a first estimate of this star's distance the Hipparcos parallax is used to give $d = 302$ pc. Then, $E(B - V)(l, b)_\infty = 0.495$ mag, and so $E(B - V) = 0.139$ mag.

However, Arce & Goodman (1999) caution that the Schlegel et al. (1998) reddening maps overestimate the reddening values when the color excess $E(B\text{-}V)$ is more than ≈ 0.15 mag. Hence, according to Schuster et al. (2004), a slight revision of the reddening estimate has been adopted via an equation, $E(B - V)_A = 0.10 + 0.65(E(B - V) - 0.10)$ when $E(B - V) > 0.10$, otherwise $E(B - V)_A = E(B - V)$, where $E(B - V)_A$ indicates the adopted reddening estimate. This leads to $E(B - V)_A = 0.125$ mag for HD 207331, and $E(b - y) = 0.093$ mag, from the relation $E(B - V) = 1.35E(b - y)$ of Crawford (1975). Then, according to the relations: $V_0 = V - 4.3E(b - y)$, $m_0 = m_1 + 0.3E(b - y)$, and $c_0 = c_1 - 0.2E(b - y)$ (Strömgren 1966, Crawford 1975), a first estimate for the dereddened photometry of HD 207331 is obtained: $(V_0, (b - y)_0, m_0, c_0) = (7.930, 0.032, 0.178, 0.999)$. From Table II of Crawford (1978) this leads to $\beta \approx 2.852$, and from his Table V, $M_V \approx 1.044$ mag, assum-

Table 2: Detected frequencies in the current study. S/N is the signal-to-noise ratio in amplitude. Var. is the percent of the total variance explained by each peak. ν_a is a possible oscillation frequency.

	Freq. (μHz)	A (mmag)	φ (rad)	S/N	Var. %
ν_1	244.7	6.0	-1.2	4.5	52
ν_2	311.1	2.9	0.9	3.7	20
ν_a	250.0	2.5	0.1	2.0	17

ing that HD 207331 is of luminosity class V, leading to an improved distance of $d = 238$ pc. This process has then been iterated twice more to a consistent solution: $E(B-V)_A = 0.110$ mag, $E(b-y) = 0.081$ mag, $(V_0, (b-y)_0, m_0, c_0) = (7.980, 0.044, 0.174, 1.002)$, $\beta = 2.854$, $M_V = 1.058$ mag, and $d = 242$ pc. These final intrinsic colors of HD 207331 are very consistent with the spectral type of A0 given by SIMBAD, according to Table II of Crawford (1978).

At the effective temperature of an A0-type star, the line blanketing due to metallic atomic absorption lines is negligible, and so the m_1 index of the Strömgren system is no longer useful for providing stellar metallicity measures. So, for HD 207331 a solar metallicity has been assumed for the following analyses.

Frequency analysis

The amplitude spectra of the differential time series were obtained by means of an iterative sine wave fit (ISWF; Ponman 1981) and the software package Period04 (Lenz & Breger 2005). In both cases, the frequency peaks are obtained by applying a non-linear fit to the data. Both procedures allow to fit all the frequencies simultaneously in the magnitude domain. Since both packages yielded similar results, we present the spectral analysis only in terms of ISWF. The amplitude spectrum of the differential light curve HD 207331 − C1 is shown in the top panel of Fig. 4. The subsequent panels in the figure, from top to bottom, illustrate the prewhitening process of the frequency peaks in each amplitude spectrum. The same procedure was followed as explained in Alvarez et al. (1998). In particular, according to these authors, a peak is considered as significant when the signal-to-noise ratio in amplitude is larger than 3.7. The oscillation frequencies detected in HD 207331 are listed in Table 2. S/N is the signal-to-noise ratio after the prewhitening process. Also shown is the fraction of the total variance contributed by each peak.

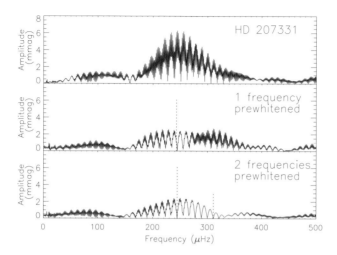

Figure 4: Amplitude spectrum of HD 207331 and the prewhitening process of the detected peaks.

The highest amplitude peak is located at 244.7 μHz $(21.1 \; \text{cd}^{-1})$, and the next significant frequency is located at 311.1 μHz $(26.9 \; \text{cd}^{-1})$. We note, however, that the second frequency is at the limit of our detection level. Given our poor window function, with only three nights of observations, it is not possible to detect more frequencies in HD 207331. Nevertheless, in the residual amplitude spectrum, after prewhitening ν_2, another peak at $\nu_a = 250 \; \mu$Hz seems to be present but with only a S/N of 2.0.

Conclusions

We have presented an analysis of CCD observations and $uvby$ differential photometry of the new δ Scuti star HD 207331, which has been found to be a multiperiodic pulsator with at least two modes of oscillations. A first estimate for the interstellar reddening has been taken from the reddening maps of Schlegel et al. (1998) via the web-page calculator of the NED database. The resulting interstellar excess is $E(b - y) = 0.081$ mag, and the dereddened photometry is consistent with the classification of an early A star.

To date, our observations represent the most extensive work on HD 207331. Beyond this, more observations, better distributed in time, are needed for an improved understanding of this interesting object.

Acknowledgments.   We would like to thank the assistance of the staff of the OAN-SPM during the observations. This paper was partially supported by Papiit IN108106, and by CONACyT project 49434-F. We thank the anonymous referee for his valuable comments which helped us to improve the manuscript.

References

Alvarez, M., Hernández, M. M., Michel, E., et al. 1998, A&A ,340, 149

Arce, H. G., & Goodman, A. A. 1999, ApJ, 512, L135

Crawford, D. L. 1975, PASP, 87, 481

Crawford, D. L. 1978, AJ, 83, 48

Crawford, D. L. 1979, AJ, 84, 1858

Lenz, P., & Breger, M. 2005, CoAst, 146, 53

Perryman, M. A. C., Lindegren, L., Kovalevsky, J., et al. 1997, A&A, 323, L49

Ponman, T. 1981, MNRAS, 196, 583

Schlegel, D. J., Finkbeiner, D. P., & Davis, M. 1998, ApJ, 500, 525

Schuster, W. J., Beers, T. C., Michel, R., et al. 2004, A&A, 422, 527

Schuster, W. J., & Nissen, P. E. 1988, A&AS, 73, 225

Strömgren, B. 1966, ARA&A, 4, 433

Comm. in Asteroseismology
Vol. 156, 2008

Modeling the Pulsating sdB Star PG 1605+072

L. van Spaandonk,[1] G. Fontaine,[2] P. Brassard,[2] and C. Aerts[1]

[1] Department of Astrophysics/IMAPP, Radboud University, PO Box 9010, 6500 GL Nijmegen, The Netherlands
[2] Département de Physique, Université de Montréal, C.P. 6128, Succ. Centre-Ville, Montréal, Québec H3C 3J7, Canada

Abstract

In this paper, we exploit new photometric data for the short period pulsating sdB star PG 1605+072 gathered with the Canada-France-Hawaii Telescope. We identify some 65 frequency components, 19 of which are due to harmonics and nonlinear combinations of larger amplitude modes. We attempt to model the 46 remaining components in terms of a collection of rotationally-split frequency components in a relatively fast rotating pulsator and, thus, determine its rotation period. To do this, we must identify a priori the central ($m = 0$) components of multiplets since the comparison is carried out with purely spherical models. We find that this a priori approach is not satisfactory and that no clear signature of a specific rotation period emerges. In the light of this result, we investigate another approach where PG 1605+072 is instead seen as a "slow" rotator in which the (numerous) low amplitude frequency components can be interpreted not as rotationally-split features but as second- and third-order harmonics and nonlinear combinations of the ten highest amplitude frequencies. This new approach can partly be justified from the observation that the light curve of PG 1605+072 does not resemble any other light curve from short period sdB stars as it shows a highly nonlinear behavior. We present some preliminary results on the basis of this assumption.

Individual Objects: PG 1605+072

Introduction

PG 1605+072 has, since its discovery by Koen et al. (1998), puzzled many scientists. Its pulsation spectrum contains over 50 frequencies, and it is commonly believed that this complex structure in the Fourier domain is due to rotational splitting in a relatively fast rotator. Indeed, high resolution spectroscopic observations have revealed significant extra broadening in narrow metal

Table 1: Photometric observations on PG 1605+072

Run	Date dd/mm/1997	Start (UT)	Number of points
cfh-054	10/06	08:23:31	525
cfh-056	11/06	08:15:56	1800
cfh-059	12/06	07:57:33	2070
cfh-062	13/06	08:17:49	1945

lines which, attributed to rotation, leads to a projected equatorial velocity of $v \sin i = 39 \, \mathrm{km \, s^{-1}}$ (see Heber et al. 1999). These characteristics are different from those commonly found in other pulsating sdB stars.

In an interesting development, Kuassivi et al. (2005) found that part of the line broadening in the spectrum of PG 1605+072 could originate from Doppler shifts (pulsational broadening) instead of a fast rotation. This could mean that the pulsation spectrum of PG 1605+072 does not contain significant frequency splitting due to rotation and that therefore another origin for its complexity needs to be found.

In this paper, we present new photometric data taken with the Canada-France-Hawaii Telescope (CFHT), as well as the results of some of our attempts to best fit the complex observed frequency spectrum of PG 1605+072 in terms of models generated by the Montréal second generation codes under the assumption that the dense spectrum originates from a fast rotation or that the star is not rotating rapidly and the spectrum has a different origin.

Observations

Photometric observations of PG 1605+072 were done during four successive nights in 1997, from June 10 to June 13, with the CFHT. For all kinds of reasons, these data had remained unexploited. Data points were taken with an integration time of 10 seconds in white light using the three-channel photometric instrument LAPOUNE. Table 1 gives some details about the four runs. The obtained light curves of the four nights can be found in Figures 1 and 2 in which one can see the complex luminosity changes as a function of time.

Frequency Analysis

For PG 1605+072, a total of 65 frequencies were found (see Table 2). The prewhitening process at different stages can be seen in Figure 3. As one can see, no peaks remain above the 3σ level in the residual of the modeled spectrum.

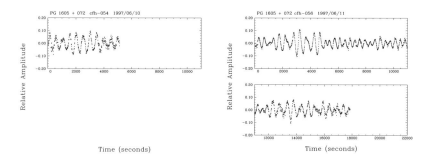

Figure 1: The lightcurve of PG 1605+072 for June 10 and 11 1997 at the Canada-France-Hawaii Telescope.

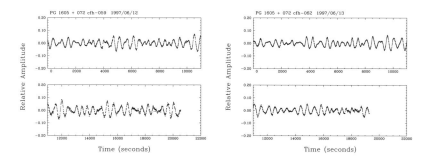

Figure 2: The lightcurve of PG 1605+072 for June 12 and 13 1997 at the Canada-France-Hawaii Telescope.

Given the relatively limited temporal resolution of our data set (our formal resolution is 3.6 μHz) in conjunction with the very dense frequency spectrum observed in PG 1605+072, some caution was necessary during the prewhitening exercise. In particular, if a peak was present around an identified mode in the data set of Kilkenny et al. (1999), that alias was chosen to participate in the process closest to the frequency from Kilkenny and company. This is because Kilkenny et al. have a much better resolution, $\sim 1/(14$ days) as compared to $\sim 1/(4$ days) in our case. From their 55 frequencies, a total of 38 could easily be located in our spectrum. Those common frequencies are indicated by asterisks in Table 2 and can be considered quite reliable. In both sets, the frequency with the largest amplitude is the same and the majority of the

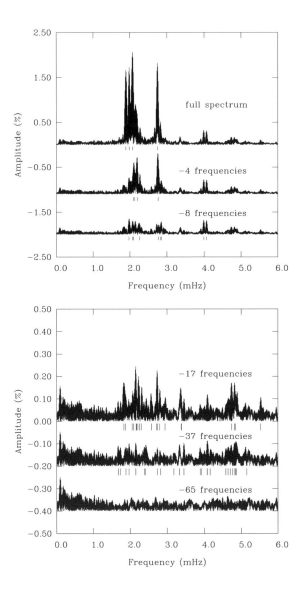

Figure 3: The prewhitening process of PG 1605+072. On the left, the full spectrum and the spectrum after subtraction of the first 4 and the first 8 frequencies. On the right, the spectrum after subtraction of the first 17 and the first 37 frequencies, and the residual spectrum after subtraction of some 65 frequencies.

Table 2: Found frequencies in the Fourier Transform of PG 1605+072

Rank n	P (s)	f (mHz)	Amplitude (%)	Rank n	P (s)	f (mHz)	Amplitude (%)
46	592.91	1.687	0.0882	5*	362.11	2.762	0.8217
52*	573.35	1.744	0.0780	16*	361.56	2.766	0.2187
20*	545.61	1.832	0.1869	27*	357.03	2.801	0.1374
24	534.61	1.871	0.1463	65*	353.73	2.827	0.5498
40	529.22	1.890	0.1463	12*	352.14	2.840	0.2683
3*	528.67	1.892	1.7856	14*	351.63	2.844	0.2600
60	518.01	1.930	0.0651	32*	339.86	2.942	0.1128
64	507.19	1.972	0.0610	61	314.30	3.182	0.0641
9*	505.82	1.977	0.3600	58	300.16	3.332	0.0874
39	505.29	1.980	0.1004	45	299.60	3.338	0.0946
4*	503.59	1.986	1.5229	26*	295.95	3.379	0.1394
28*	485.74	2.059	0.1253	37*	295.34	3.385	0.1023
1*	481.79	2.076	2.0640	42	289.69	3.452	0.0990
11*	481.71	2.076	0.2963	51	257.23	3.889	0.0801
29*	476.73	2.098	0.1244	63	254.76	3.925	0.0623
7*	475.95	2.101	0.6601	13*	246.25	4.061	0.2635
15*	475.85	2.102	0.2250	10*	250.44	3.993	0.3248
8*	475.27	2.104	0.5450	38	245.40	4.075	0.0679
46	463.44	2.158	0.0926	48	245.25	4.077	0.1008
18*	461.37	2.167	0.1980	56*	240.58	4.156	0.0724
33	459.61	2.176	0.1118	52*	218.23	4.582	0.0790
6*	454.19	2.202	0.8106	54	215.36	4.643	0.0777
19	451.84	2.213	0.1943	49	212.63	4.703	0.0838
25	443.28	2.256	0.1397	21	210.99	4.740	0.1710
17*	440.32	2.271	0.2134	55	209.58	4.771	0.0763
30*	434.30	2.303	0.1172	23*	207.54	4.818	0.1612
57*	418.00	2.392	0.0713	44	207.16	4.827	0.0984
43	412.87	2.422	0.0989	41*	206.52	4.842	0.0999
36*	387.45	2.581	0.1025	35*	206.38	4.845	1.0489
22*	368.16	2.716	0.1698	59	205.11	4.875	0.0671
62*	365.28	2.738	0.0635	50	195.14	5.125	0.0824
31*	364.72	2.741	0.1159	34	180.96	5.526	0.1071
2*	364.60	2.743	1.8283				

Table 3: The equally spaced multiplets present in the observed spectrum. Multiplets 1, 2, 3, and 4 have a spacing of ~ 90 μHz. Multiplets 5, 6, 7, and 8 are have a spacing of ~ 40 μHz.

Nr.	f_{center} mHz	m	$f \pm \Delta\nu$ mHz	$\Delta\nu$ μHz	Nr.	f_{center} mHz	m	$f \pm \Delta\nu$ mHz	$\Delta\nu$ μHz
		-1	1.744	88			-1	1.890	40
1	1.832	0	-	-	5	1.930	0		-
		1	1.930	98			1	1.972	42
		-2	1.892	184			-2	2.716	85
		-1	1.986	90			-1	2.762	39
2	2.076	0	-	-	6	2.801	0		-
		1	2.167	91			1	2.844	43
		2	2.256	180					
							-2	2.176	80
		-1	2.213	90			-1	2.213	43
3	2.303	0	-	-	7	2.256	0		-
		1	2.392	89			1	2.303	47
		-2	2.582	180			-2	1.892	85
4	2.762	0	-	-	8	1.977	0		-
		2	2.942	180			2	2.059	82
							3	2.102	125

10 highest amplitude frequencies is the same, although not in the same order. The other frequencies in Table 2 are less reliable, but we nevertheless believe them to be real although possibly less accurate. Given the limitations of current models in reproducing the observed frequencies at their observational precision, this question of frequency accuracy is of secondary importance in the context of this exploratory work.

Models

Using the list of frequencies given in Table 2, we tried to construct acceptable seismic models for PG 1605+072 using the Montréal second generation model building codes (see, e.g., Brassard et al. 2001). We searched parameter space in the ranges $T_{eff} = 31,500-33,500$ K (with a resolution of 10 K), $\log g = 5.200-5.300(0.001)$, $M_*/M_\odot = 0.300-0.800$ (0.001), and $\log q(H)$ from -3.00 to -6.00 in steps of 0.01. The search domains for the effective temperature and the surface gravity were defined by independent spectroscopic constraints.

Fast Rotation Models

Kawaler (1999) was the first to try to find an asteroseismological model for PG 1605+072. He only used the five frequencies with the highest amplitudes within the dense Fourier spectrum reported by Kilkenny et al. containing at least 55 frequencies, and found that three of these frequencies could possibly make up a rotational triplet if the velocity at the equator of PG 1605+072 is 130 km s^{-1}. Taken at face value, and assuming log $g = 5.25$ and a representative mass of $0.5\,M_\odot$ for PG 1605+072, this implies a rotation period of about 2.6 h. This is quite short for a single star such as PG 1605+072. At the same time, this also implies that the temporal resolution achieved during our 4-night observing run is amply sufficient to resolve rotationally-split multiplets, if present (not withstanding the difficulty of overlapping frequency components in the dense spectrum as briefly alluded to above). Subsequent spectral analysis by Heber et al. (1999) indicated that PG 1605+072 could indeed be a fast rotator, with a projected rotational velocity of $V \sin i = 39$ km s^{-1}, a rather high value by sdB standards. Starting from these findings, we tried to find a rotation period for PG 1605+072 by looking for nearly equally spaced components within frequency multiplets caused by the fast rotation as expected from first-order perturbation theory. The different multiplets found are listed in Table 3. Next, we retained what appeared to us as the central ($m = 0$) components of the multiplets for detailed comparison with the predictions of spherical models in the same spirit as in Brassard et al. (2001).

Model 1

In this first model, a reference one, we assume no rotation at all, and we try to fit all observed frequencies as independent modes with different values of k and l. It should be pointed out, however, that in the original list of 65 frequencies only 46 can be considered a priori as independent modes, the others (those with periods less than 260 s) being harmonics and nonlinear combinations of the highest amplitude modes. Hence, we tried to fit only 46 frequencies. As expected for this reference model, the best fit is really not good, showing a rather large goodness-of-fit value of $\chi^2 = 741$. This indicates that our original assumption that all 46 frequencies are modes with different individual values of k and l is incorrect. For a graphical representation of the model fit, see Figure 4.

Model 2

In this second model, we removed the 10 frequency components associated with the m components of the four multiplets characterized by a common spacing of 90 μHz in Table 3. We thus retained only the $m = 0$ components of these multiplets, limiting our list of retained frequencies to 36. Again, we searched in parameter space for the best fit model. We found that the resulting merit function for that optimal model is still very high at $\chi^2 = 349$. Taking into account the reduced number of frequencies to fit, this is no significant improvement over the previous case. See Figure 4 for a graphical representation of the model fit.

Model 3

Our third search exercise is similar to our previous attempt, except that we removed 10 other frequencies, this time associated with the four multiplets in Table 3 showing a possible common spacing of 40 μHz, the other possibility for such splitting in our original list of 46 frequencies. Hence, we tried again to fit 36 frequencies, but 10 of which are different from before. We then found a significant improvement in the merit function over the previous cases, a value of $\chi^2 = 215$, but this still remains an unacceptable fit as can be seen in Figure 4. The observed distribution of periods is simply not well reproduced by our approach.

If PG 1605+072 is rotating fast enough to split the modes into equally spaced multiplets, the subtraction of these components should significantly improve the best fit to spherical models. As this is not the case, one could conclude that perhaps PG 1605+072 is rotating so rapidly that second-order solid-body rotation or differential rotation effects would set in, resulting in non-equally spaced multiplets for which the search is beyond the scope of this work. Alternatively, one could instead conclude that perhaps PG 1605+072 is, in fact, rotating slowly, and rotational multiplets can no longer be found with our frequency resolution. At this stage, we prefer to think that the weakness of our approach rests currently with our a priori identification of the $m = 0$ components of the multiplets. We suggest that the period-matching exercise in parameter space should be redone in the future, but using models that specifically include rotation so that no a priori identification needs to be made. This would allow all the 46 observed frequencies to be treated on the same footing. An example of such an approach has been presented recently by Van Grootel et al. (2008) for the short-period pulsating sdB star Feige 48.

Slow Rotation Models

The idea that PG 1605+072 may not be, after all, a fast rotator comes from the work of Kuassivi et al. (2005) who showed that pulsational broadening could account for up to 34 km s^{-1} (twice the measured fluid displacement velocity of 17 km s^{-1}) in terms of line broadening in that star. If true, it may be that the rotation period of PG 1605+072 is rather long. Of course, if this is the case, another mechanism needs to be found for the large number of visible modes in the spectrum. Comparing the looks of the light curves in Figure 1 and

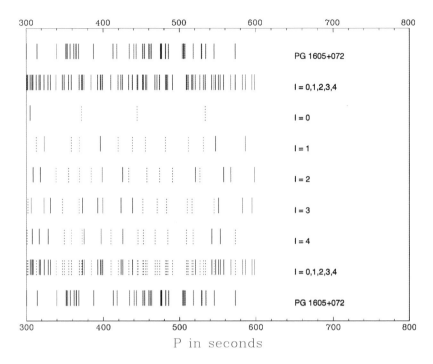

Figure 4: Graphical representations of the best-fit models for a fast rotating PG 1605+072: Model 1

The first line of each graph is the observed spectrum of PG 1605+072, the second line represents the spectrum for the best fit model, the next four lines split this spectrum into the different l-values, and the last two lines are similar to the first two lines. Dashed lines represent the modes in the modeled spectrum identified with modes in the observed spectrum.

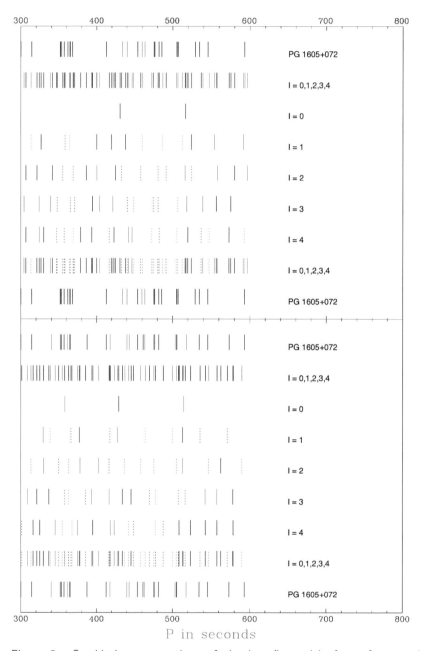

Figure 5: Graphical representations of the best-fit models for a fast rotating PG 1605+072. The configuration in the figure is the same as in Figure 4. Top panel: Model 2. Bottom panel: Model 3.

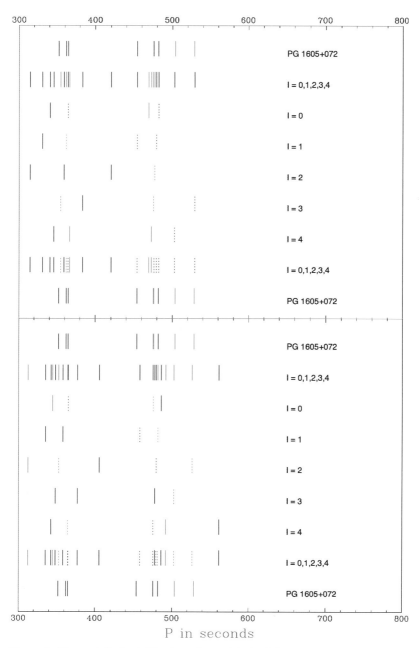

Figure 6: Figure 6 (below) The best fit models for PG 1605+072, including only the ten highest amplitude modes. The configuration in the figure is the same as in Fig. 4. Top panel: Model 4. Bottom panel: Model 5.

Figure 2 with other sdB light curves suggests that, for PG 1605+072, a highly nonlinear oscillation process could be at the origin of the many frequencies in the spectrum. The result of this is that a large number of 2nd- and 3rd-order harmonics and cross frequencies can be invoked to make up the rich spectrum of PG 1605+072. In fact, among our list of 65 frequencies, it only takes the ones with the ten highest amplitudes to construct the remaining 55 frequencies with second- or third-order combinations within our resolution.

Model 4

Using these 10 modes only, another search in parameter space results in a best fit model defined by $T_{\text{eff}} = 32\,300$ K, $M = 0.707 M_{\odot}$, $\log g = 5.248$, and $\log q(\text{H}) = -5.78$. The merit function for this model is $\chi^2 = 15.3$. Not only is the formal fit better than in the previous models (taking into account the much reduced number of observed frequencies in the fit), but the distribution of the modeled modes is now very similar to the distribution of the observed modes as can be seen in Figure 5. Furthermore, the highest amplitude mode is either an $l = 0$ or an $l = 1$ mode, as can be expected. Although the mass of this model is rather high (and therefore the hydrogen envelope rather thin), it is still within the predicted mass range by Dorman et al. (1993), Han et al. (2002), and Han et al. (2003).

Model 5

Another good fit for the 10 highest amplitude modes is found at $T_{\text{eff}} = 32\,300\,K$, $M = 0.561 M_{\odot}$, $\log g = 5.217$ and $\log q(\text{H}) = -6.22$ (see Figure 5). The fit for this model is slightly worse than for Model 4 at $\chi^2 = 24.3$, but still noteworthy because of the lower mass. The l value for the highest amplitude mode in this model is either 1 or 2.

Conclusion

New data for the pulsating sdB star PG 1605+072 have been presented. After analysis, a pulsation spectrum containing at least 65 frequencies emerges. During the prewhitening process, we used, when appropriate, the mode identification of Kilkenny et al. (1999), and we recovered 38 of their frequencies.

Under the assumption that the dense spectrum is caused by a relatively fast rotation of the star which produces a splitting of modes into equally spaced multiplet components, we found two possible frequency spacings at $\sim 90\ \mu$Hz and $\sim 40\mu$Hz. The subtraction of the m components from the spectrum did not improve the quality of the fit between the observed values and the model frequency spectrum, however, as could have been expected. We suspect that this failure is caused by our a priori identification of the $m = 0$ components, and we suggest that the period-matching exercise in parameter space should be redone but, this time, with models incorporating rotation explicitly.

If the dense spectrum is assumed instead to consist of a large number of second- and third-order harmonics and cross frequencies, and only the highest amplitude modes are genuine modes, we then find two acceptable models at $M_* = 0.561 M_\odot$ and $M_* = 0.707 M_\odot$. Further investigations along this line of nonlinear frequency peaks should certainly be pursued in more detail than presented here, particularly in view of the unique, highly nonlinear light curve observed for PG 1605+072.

References

Brassard, P., Fontaine, G., Billères, M., et al. 2001, ApJ, 563, 1013

Dorman, B., Rood, R.T, & O'Connell, R.W. 1993, ApJ, 419, 596

Han, Z., Podsiadlowski, P., Maxted, P.F.L., et al. 2002, MNRAS, 336, 449

Han, Z., Podsiadlowski, P., Maxted, P.F.L., & Marsh, T.R. 2003, MNRAS, 341, 669

Heber, U., Reid, I.N, & Werner, K. 1999, 343, L25

Kawaler, S.D. 1999, ASPC, 169, 158

Kilkenny, D., Koen, C., O'Donoghue, D., et al. 1999, MNRAS, 303, 525

Kuassivi, Bonanno, A., & Ferlet, R. 2005, A&A, 442, 1015

Koen, C., O'Donoghue, D., Kilkenny, D., et al. 1998, MNRAS, 296, 317

Van Grootel, V., Charpinet, S., Fontaine, G., & Brassard, P. 2008, A&A, 483, 875

Comm. in Asteroseismology
Vol. 156, 2008

Data Reduction pipeline for MOST Guide Stars and Application to two Observing Runs

M. Hareter[1], P. Reegen[1], R. Kuschnig[1], W. W. Weiss[1], J. M. Matthews[2],
S. M. Rucinski[3], D. B. Guenther[4], A. F. J. Moffat[5], D. Sasselov[6],
G. A. H. Walker[2]

[1] Institut für Astronomie, Türkenschanzstrasse 17, A-1180 Vienna, Austria
[2] Department of Physics & Astronomy, University of British Columbia,
6224 Agricultural Road, Vancouver, B.C., V6T 1Z1, Canada
[3] Department of Astronomy, University of Toronto, 50 St.George Street,
Toronto, Ontario, M5S 3H4, Canada
[4] Department of Astronomy and Physics, St. Mary's University,
Halifax, Nova Scotia, NS B3H 3C3, Canada
[5] Département de physique, Université de Montréal, Montréal,
Québec, QC H3C 3J7, Canada
[6] Harvard-Smithsonian Center for Astrophysics, Cambridge,
Massachusetts, MA 02138, USA

Abstract

A Data Reduction pipeline for MOST[1] Guide Star data is presented together
with the results obtained for two observing runs in 2004 (i.e. κ^1 Ceti 2004
and HR 1217 2004 runs) containing four and seven Guide Stars respectively.
Among these Guide Stars, four are clearly variable with only one known before
the MOST observations: the long period variable HD 24338 (M2III). The data
reduction relies on the decorrelation technique employed by Reegen et al. (2006)
for their data reduction pipeline of MOST Fabry targets. The main difference is
that the MOST Guide Star data include no background information. A coarse
on-board background subtraction is performed, but leaves considerable residual stray light in the data, which is subject to a more sophisticated reduction

---

[1] Based on data from the MOST (Microvariability & Oscillation of STars) satellite,
a Canadian Space Agency mission jointly operated by Dynacon, Inc., the University of
Toronto Institute of Aerospace Studies, and the University of British Columbia, with
assistance from the University of Vienna, Austria.

technique. Since at least four stars are observed simultaneously, common features of the light curves can be recognized and removed. Among different data reduction methods, the decorrelation technique is more versatile and often has better results than differential photometry or data smoothing.

Individual Objects: HD 20884, HD 20790, HD 24338, HD 24217

Introduction

MOST (Microvariability and Oscillations of STars) is a Canadian microsatellite space mission, which carries a 15-cm Rumak-Maksutov telescope. For a full description before and after launch see Walker et al. (2003) and Matthews (2004). Three different photometric modes are available simultaneously:

(a) Fabry imaging: the telescope entrance pupil is imaged as an annulus of 40 pixels diameter, enhancing the total signal and reducing the influence of satellite jitter. Reegen et al. (2006) developed a reduction pipeline for data obtained in this mode, resolving linear correlations between target and background pixels.

(b) Direct imaging: stars are imaged directly, i.e. the light does not pass through one of the Fabry lenses. For this mode, 20×20 pixel subrasters are used and the images are stored as extentions in the FITS headers. Rowe et al. (2006) and Huber & Reegen (2008) developed direct imaging data reduction pipelines independently, the latter relying on the decorrelation technique.

(c) Guide Star photometry: this mode of photometry was originally intended for guiding only, but the data quality is sufficiently high to be used for science. A data reduction pipeline for this mode is presented here. In early 2006 the MOST Guiding CCD, which was used for gathering the Guide Star data, failed due to a charged particle hit. Since this event the Science CCD took over its function. Originally, the reduction method presented here was developed for the Guiding CCD. Since the data format did not change, it is fully applicable for Guide Star data from the Science CCD.

MOST light curves suffer from stray light from the bright Earth. A periodic non-sinusoidal modulation of the light curves is present in any observed data set. A detailed study of the sources of stray light and other instrumental effects can be found in Reegen et al (2006). To mitigate the stray light and other instrumental signal, the decorrelation method was employed for the Guide Star data.

The data reduction routine consists of two IDL-based programs, one for data extraction and one for the data reduction. The frequency analysis was done using SigSpec (Reegen 2007). Meanwhile more than 1000 stars have been observed. In this paper, we denote Guide Stars by the lower case letter g followed by a number. For the sake of convenience, we modify the Julian Date (JD) to JDM = JD - 2 451 545 (epoch 2000.0). In the follwing sections, the data format of MOST observations is described, three approaches for data reduction are discussed and the results of two runs using the decorrelation technique are described.

Data Format and On-board Background Subtraction

MOST has two Science Data Stream (SDS) formats: one contains compressed images of the primary science target (SDS1) and one contains a 2×2 pixel binned image (SDS2). Both SDS formats contain information on the Guide Stars within the FITS headers. One FITS file usually contains one exposure of the Fabry image, the direct images (20×20 pixel subrasters) as extensions and the Guide Star data as header entries. No Guide Star images are available and the only information available is the intensity and the integration time, as shown by the following example:

```
...
JD_OBS      1748.2233094  / Julian Date - 2,451,545
NUM_GS      4             / Number of guide stars
GS_I0000    144114        / Intensity of guide star number 0
GS_T0000    26500         / Integration time for guide star in milliseconds
GS_I0001    35430         / Intensity of guide star number 1
GS_T0001    26500         / Integration time for guide star no. 1
...
```

In later FITS header versions, the Heliocentric Julian Date was implemented and labeled JD_HEL. At least four Guide Stars must be observed simultaneously for pointing requirements. Due to the limited downlink capacity of MOST, the Guide Star information is processed and compressed on board the satellite. For each Guide Star, a 20×20 pixel (or $1' \times 1'$) raster is used. Short exposure times are required due to pointing requirements, hence image stacking is used. Individual frames to be stacked are hereafter referred to as subexposures. The integration time of a subexposure is typically between 0.3 and 3.0 seconds. The smaller the integration time, the higher the sampling rate and the better the pointing. On the other hand, the longer the integration time, the better the photometric precision.

The background is determined as the mean value of the first and the last row of the individual subraster. This mean value is referred to as threshold. Inten-

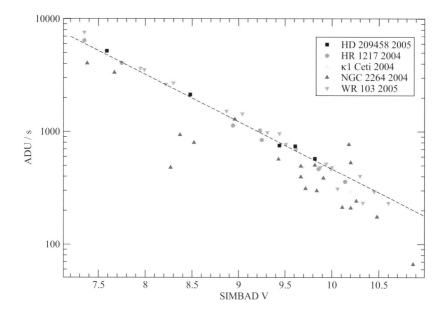

Figure 1: V magnitude versus count rate (ADU/s). Guide Star data of five observing runs are shown. The dashed line corresponds to the expected count rate for a given magnitude.

sities exceeding the threshold are considered stellar signal. Only the intensities of these pixels are summed up, subtracting the individual threshold from each subexposure. This corresponds to an aperture photometry, where the aperture is determined dynamically. The intensity of a stacked exposure (e.g GS_I0000 listed above) is obtained by summing up the intensities of all subexposures. Since fainter stars have fewer illuminated pixels, this kind of aperture photometry produces systematically lower intensities for fainter stars than expected, which is illustrated in Fig. 1. Mean ADU values for individual stars are plotted vs V magnitude according to SIMBAD. Since MOST has a custom broadband filter, those values are not exactly comparable, but to a sufficient accuracy for our purpose. The dashed line corresponds to the expected ADU value for a Guide Star with given magnitude m_V. Fainter stars are systematically below the line.

Unfortunately, the threshold is not stored in the FITS header. Hence no information on the background is available. This limits the precision of the Guide Star photometry, but still the data are of high quality and time coverage.

Data reduction Techniques

Decorrelation

For data reduction, we used the approach developed for science targets observed in Fabry Imaging mode (Reegen et al. 2006). This method resolves linear correlations between the intensities of target pixels and those of background pixels. Since no background information is available for Guide Stars, the light curves of non-variable Guide Stars are used instead. The observed light curves of the Guide Stars are classified as variables and non-variables by a quick-look analysis. Correlation coefficients between a particular variable star's light curve and all non-variable light curves are calculated.

Subsequently, the comparison light curves are sorted by descending correlation coefficients. Then the linear regressions of the variable star's light curve in relation to the light curves of comparison stars are subtracted from the variable's intensities. The rank of correlation coefficients determines the order in which non-variable light curves are picked. All light curves are decorrelated vs. the light curves of all reference stars according to their ranking.

The number of decorrelations is the number of constant reference stars in the sample. Resolving linear correlations between light curves mitigates all Fourier amplitudes of a variable star rather than only the unwanted ones. In the case of Guide Star photometry, the amount of decorrelations performed is small enough to neglect this issue.

Fig. 2 represents the intensity vs intensity plot of the two light curves of non-variable Guide Stars g6 and g1 and the linear regression which is subtracted subsequently. Correcting for linear regression causes common signal (e.g. orbital variations) to be mitigated. Since the pattern (see Fig. 2) is generally more complicated than the linear approximation (Fig. 9 in Reegen et al. 2006), stepwise multiple linear regression is employed: different CCD positions are exposed to different stray light conditions, and correcting for more than one reference star permits to cover a greater variety of contamination patterns.

Fig. 3 shows the effect of decorrelation in a sample with many constant reference stars. The periodic modulation at the MOST orbit frequency at $14.19\,\mathrm{d}^{-1}$ is reduced by a factor of about 10. The presented part of the light curve corresponds to a δ Scuti star observed by MOST simultaneously with Procyon in 2007. The peaks near $25\,\mathrm{d}^{-1}$ indicate intrinsic signal, which is practically not affected by the decorrelation method. This reduction of instrumental signal is not due to outlier rejection, because the term "raw data" is used here in the sense of outlier corrected data before decorrelation.

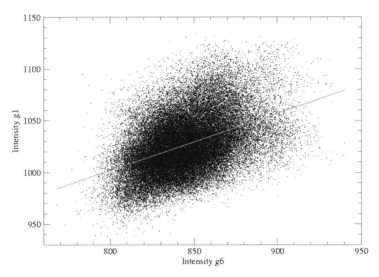

Figure 2: Intensity vs. intensity plot of two Guide Stars of the HR 1217 2004 run. The solid line shows the linear regression.

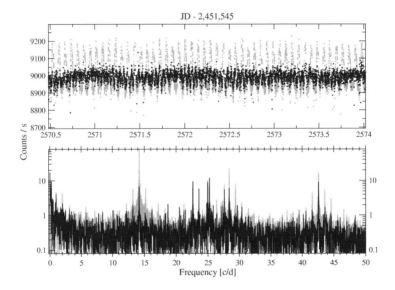

Figure 3: Unprocessed data (gray) compared to decorrelated data (black). In this particular observing run 12 non-variable Guide Stars were available. The decorrelation technique significantly reduced the residual power near the MOST orbit frequency $(14.19 \ d^{-1})$ and near its harmonics. The scale on the y-axis is logarithmic for better visibility.

Limitations

There are a few limitations to the decorrelation method. When applied to MOST Fabry data, the background pixels are well-defined and isolated from the stellar signal. The MOST Guide Star photometry does not provide independent information in the form of background pixels on the CCD, hence presumably constant reference stars must be used instead. The correlation of background pixels to target pixels is generally better than the correlation between stellar light curves.

The decorrelation method cannot identify a variable star independently. Using a variable as a reference introduces spurious signal into all the stars in the sample; thus, a variable may not be discovered. Fig. 4 shows the effect of the decorrelation technique, if a variable reference star is used accidentally. The intrinsic signal of g1 is introduced into all other constant reference stars (g2 vs. g1 is shown as an example). Comparing the resulting frequencies after a SigSpec analysis of the individual stars, the pulsation signal of g1 may be misinterpreted as common instrumental periodicities.

A further issue is that if the stray light is extremely high or if the targets have been observed only during orbital phases of maximum stray light (for example switch targets), the decorrelation method significantly increases the time domain point-to-point scatter. For those targets, the decorrelation technique may not be an appropriate approach.

If the stray light distribution is extremely inhomogeneous on the CCD, Guide Stars show very different light curves. In this case, Guide Stars close to the variable should be used rather than stars showing very different stray light behavior.

Due to the rejection of outliers, the spectral window may become worse. Outliers produced at random and uniformly distributed in time will not affect the spectral window, but cosmic ray impacts occur more frequently during South Atlantic Anomaly passages of the spacecraft, and their rejection will introduce aliasing at the orbital frequency of the satellite. However, these peaks will show up in the spectral window and are easy to identify also in the Fourier spectrum of a light curve.

Other Approaches

Differential Photometry

The three-star differential photometry may be also applied to MOST Guide stars. However, the comparison stars have to be selected carefully for two reasons: Obviously, they may be variable and the stray light behavior may be very different at different CCD positions. Stray light may also move across the

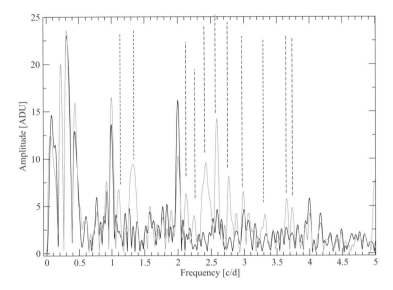

Figure 4: Fourier amplitude spectra for Guide Star g2 before (solid black line) and after (solid gray line) decorrelation vs. g1. The dashed lines mark the frequencies intrinsic to g1 which were transferred to g2 by the decorrelation.

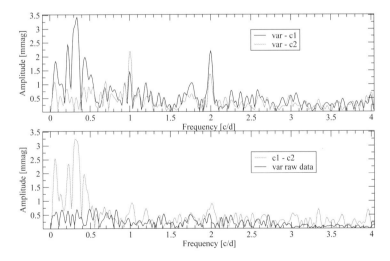

Figure 5: Application of the three star photometry to the data of the bright star g0. Top panel shows var - c1 and var - c2, bottom panel shows c1 - c2 and the Fourier spectrum of the unprocessed data. The peaks at $0.3\,d^{-1}$ may be due to intrinsic variablity of c1.

CCD, which means that stray light influences light curves of stars at different positions at different times. This effect complicates the differential photometry. In many cases, there are no appropriate comparison stars available. For example in the κ^1 Ceti 2004 run the two stars which are classified as non-variable are too faint. Fig. 4 shows the application of the three star photometry to the data of g0 (var) in the κ^1 Ceti 2004 run, where the two faint stars are used as comparison stars (c1 and c2). The pulsation signal of the bright K giant is lost in the noise introduced by the faint comparison stars. In this run, the differential photometry cannot be applied.

Another approach may be to calculate a mean light curve of the constant stars and subtract this averaged light curve from the variable star data. The problems mentioned above have to be taken into account also in this case.

Data Smoothing using Periodic Filter Functions

In the time domain, filtering may be considered the convolution of a response function $\rho(\tau)$ and the observable $x(t)$,

$$x_F(t) = \rho * x = \int_{-\infty}^{\infty} d\tau \, \rho(\tau) \, x(t + \tau) \ . \tag{1}$$

The filtered observable is then $x - x_F$.

The Fourier transform $\widetilde{x_F}$ is given by the product of the transformed response function, $\tilde{\rho}$, and the transformed observable, $\tilde{x}$, according to

$$\widetilde{x_F}(\omega) = \tilde{\rho}(\omega) \, \tilde{x}(\omega) \ . \tag{2}$$

For convenience, we use the *transformed filter function* (TFF)

$$\Psi(\omega) = 1 - \tilde{\rho}(\omega) \tag{3}$$

rather than the transformed response function $\tilde{\rho}(\omega)$.

The most elementary application of a periodic filter in order to correct for unwanted periodicities in time series data is a set of equidistant rectangular profiles of width δ. The distance of consecutive profiles in time shall be denoted Δ. In the present case, Δ is equal to the orbital period of MOST. Furthermore, we note that due to symmetry, the number of such rectangular profiles will always be odd, say $2N + 1$. The number of rectangles determines the *time bandwidth* of the filter, whereas the orbital phase interval mitigated by the filter shall be called the *phase bandwidth* and is adjusted by δ. Given these specifications, the TFF evaluates to

$$\Psi(\omega) = 1 - \left| \frac{2}{(2N+1)\,\omega\delta} \left(\sin\omega\frac{\delta}{2} + 2\sum_{n=1}^{N} \cos n\omega\Delta \, \sin\omega\frac{\delta}{2} \right) \right| \ . \tag{4}$$

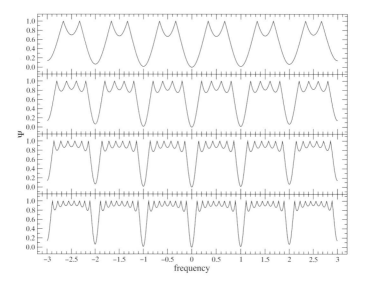

Figure 6: TFFs of periodc rectangular time-domain filters for 3, 5, 7, and 9 rectangles (*top* to *bottom*), i. e., $N = 1$, 2, 3, and 4, respectively. All panels refer to $\Delta = 1$, $\delta = 0.1$.

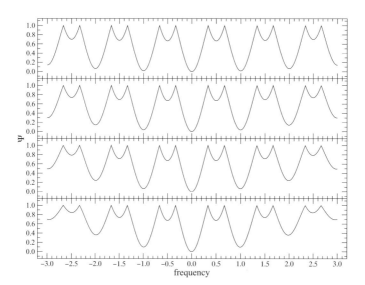

Figure 7: TFFs of periodc rectangular time-domain filters for phase bandwidths $\delta = 0.1$, 0.15, 0.2, and 0.25 (*top* to *bottom*). All panels refer to three rectangular profiles (i. e., $N = 1$) with $\Delta = 1$.

Figs. 6, 6 display the TFFs for different values of N and δ, respectively. Setting $\Delta = 1$ provides the frequency to be in units of orbital frequency. Thus, peaks reaching zero or values close to zero in the Figures refer to the vicinity of the orbital frequency of the spacecraft (and harmonics) and illustrate the efficiency of the filter. On the other hand, there are also periodicities in between these major peaks, which are unwanted, because they introduce systematic over- and under-estimation of selected frequency regions. This is a principal and serious disadvantage of periodic time-domain filters, which can be overcome by more sophisticated approaches only to a limited extent.

Fig. 6 illustrates the influence of the time bandwidth on the TFF: the TFFs for 3, 5, 7, and 9 rectangular profiles ($N = 1$, 2, 3, and 4, correspondingly) are displayed, all for a phase bandwidth $\delta = 0.1$. Employing more rectangular profiles (i. e. increasing the time bandwidth) narrows the major peaks of the TFF around the orbital frequency and its harmonics and increases the number of minor peaks in between. The amplitude of these minor peaks is also slightly reduced.

Fig. 6 contains four panels associated to the TFFs for phase bandwidths $\delta = 0.1$, 0.15, 0.2, and 0.25, all for three rectangular profiles (i. e., $N = 1$). The phase bandwidth δ influences the effect of the filter on orbit harmonics: narrow-phase-band filters are more efficient in this respect.

However, the problem that periodic time-domain filters affect the entire frequency range and consequently also stellar signal rather than only instrumental pseudo-periodicities, persists independently of the parameter settings. As an example, the light curve of HD 20790 (Guide Star g1 of κ^1 Cet) is filtered with $\Delta = 101.4$ min, $N = 5$, and $\delta = 0.1$. Fig. displays the DFT amplitude spectra of the two light curves (raw and filtered) in the top panel. The filtered spectrum (gray curve) shows the mitigated orbit-related peaks, but their $1\,\mathrm{d}^{-1}$ sidelobes, which represent modulations in the stray light contamination, remain practically unaffected. The bottom panel contains the TFF, as obtained from the amplitude spectra and clearly visualizes the effect of systematic over- and under-estimation of selected frequency regions. Unfortunately, the peak amplitudes around $1\,\mathrm{d}^{-1}$, which are of particular interest for γ Dor variables, are massively distorted.

Application to two Fields and Results

Analysis of the Guide Stars of the κ^1 Ceti 2004 Observing Run

The final reduction of the Guide Stars was performed using the decorrelation method described above. Table 1 gives the observation details of both runs. The stray light conditions in these observations were optimal, hence the raw data are of extraordinary quality compared to many other runs. Fig. 9 compares the raw data to the 3-σ clipped data and the reduced data of g1 as an

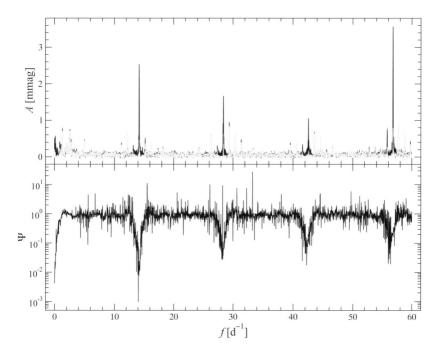

Figure 8: *Top:* Amplitude spectra of the HD 20790 Guide Star photometry. The *black* curve refers to the raw data, the *gray* curve to the light curve after application of a periodic rectangular time-domain filter with $\Delta = 101.4$ min, $N = 5$, and $\delta = 0.1$. *Bottom:* TFF derived from the amplitude spectra in the top panel.

example. The spectral window before and after processing the data are shown in Fig. 10, Fig. 11 shows the Fourier spectra of g1. The gray color refers to the 3-σ clipped data, and the Fourier spectrum of the decorrelated data is represented by the black graph. The sidelobes of the MOST orbit frequency and its harmonics are mitigated by decorrelation. However, the orbit peak at $14.19\,d^{-1}$ is increased significantly from 2.4 mmag in the raw data to 3.1 mmag in the decorrelated data. The first harmonic at $28.38\,d^{-1}$ increases slightly, whereas higher overtones are damped.

This observing run contains two variables and two rather faint comparison stars, where one (g2) might also be a long period variable, because the Fourier spectrum contains a power excess in the lowest frequency regime. Despite this fact, this star was also used for the decorrelation technique, which affects only the frequency region below $0.5\,d^{-1}$. Power in this frequency region has been ignored. The variables are considerably brighter than the two comparison stars, hence the decorrelation technique does not decrease the instrumental

Run	κ^1 Ceti	HR 1217 run
Date of the observations (2004)	Oct 15 to Nov 4	Nov 5 to Dec 4
Length of final dataset [d]	19.63	29.02
Integration time [s]	26.25	26.4
Sampling time [s]	35	35
# of raw data-points	48 869	70 066
# of data-points after reduction	44 357	64 698

Table 1: Details of the MOST photometry of the κ^1 Ceti 2004 and HR 1217 runs

ID	GSC-ID	HD / BD	SpT.	V [mag]	parallax [mas]
g0	00060-00509	HD 20884	K2III	7.47	5.76 ±0.57
g1	00060-01499	HD 20790	F4IV	8.9	6.70 ±0.86
g2	00060-00969	-	F8	10.2	-
g3	00060-00845	BD +03 459	K2III	9.7	-

Table 2: Information on the κ^1 Ceti Guide Stars including the new HIPPARCOS parallaxes (van Leeuwen 2007).

ID	Standard deviation [mmag]	Point-to-point scatter [mmag]
g0 raw	5.754	3.130
g0 reduced	5.582	3.000
g1 raw	9.740	8.145
g1 reduced	7.601	6.472

Table 3: Improvement in terms of standard deviation and point-to-point scatter of the two variable stars among the κ^1 Ceti Guide Stars by decorrelation.

artifacts significantly. However, the RMS deviation of the data sets, which are important for the frequency analysis, were decreased: for g0 only 3% reduction was achieved, for g1 22%. The results of the frequency analysis using SigSpec were compared for the four light curves. Only frequencies that are unique for the data sets, within the frequency error of ± 0.013 d^{-1} or $1/(4T)$, where T denotes the length of the dataset, were accepted for each star (Kallinger et al. 2008b).

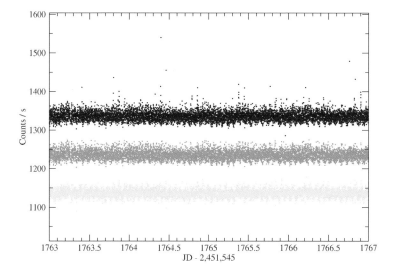

Figure 9: Comparison of the g1 light curves: raw data (top), 3-σ clipped raw light curve (middle) and decorrelated data (bottom). A subsample of four days is shown with vertical offset for the three graphs, for better visibility.

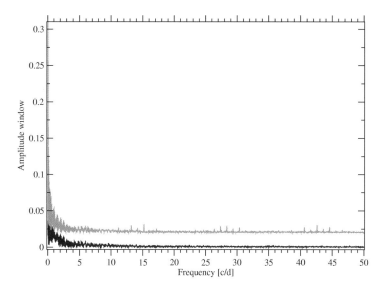

Figure 10: Comparison of the spectral windows for the raw data (black line) and the processed data (gray line, offset 0.02 for visibility). The spectral window changes marginally due to periodic rejection of outliers, hence aliasing plays a minor role.

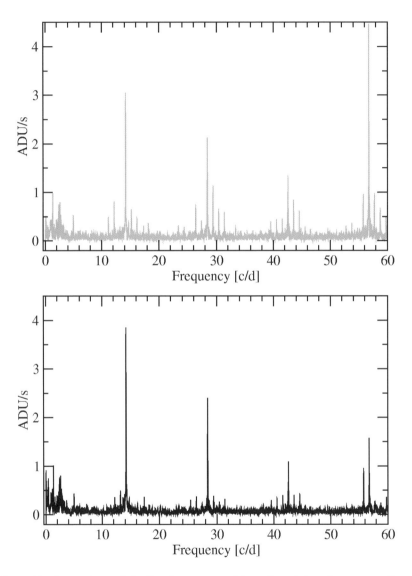

Figure 11: Comparison of the Fourier spectra of 3-σ clipped data (upper panel) and decorrelated data (bottom panel). The sidelobes of the MOST orbit frequency and also the higher order harmonics are reduced.

Results of the Guide Stars of the κ^1 Ceti 2004 Observing Run

The red giant HD 20884

The final light curves for both stars are shown in Fig. 12 and the Fourier spectrum in the range from 0 to 6 d^{-1} is displayed in Fig. 13. Kallinger et al. (2008a) found evidence for radial and nonradial pulsations in this red giant and drew conclusions concerning the mode lifetime. The authors also state that the model which fits the frequencies best is too luminous compared to the HIPPARCOS parallax published in 1997 (ESA 1997).

The new reduction of the HIPPARCOS data was published by van Leeuwen (2007) and the new parallax puts the star at a larger distance. Hence, the star is more luminous. The absolute magnitude M_V was estimated employing $m_V - M_V = 5 \log r - 5$, where r is the distance in pc. The luminosity is estimated using $L/L_\odot = 79.43 \times 10^{(-0.4(M_V + BC))}$ (Kallinger et al. 2008a). Taking the parallax error and the bolometric correction into account, the bolometric luminosity of HD 20884 is estimated between 40 and 60 $L_\odot$ which is still consistent with the value found by those authors.

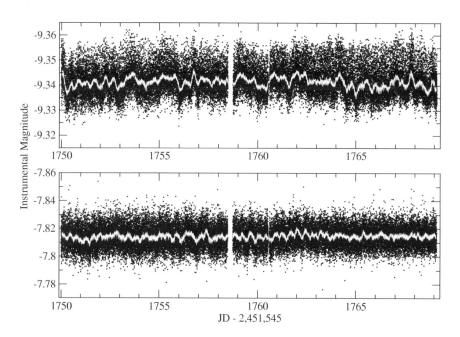

Figure 12: Final light curves of g0 (HD 20884, upper panel) and g1 (HD 20790, lower panel). Moving averages over 200 data points are overplotted.

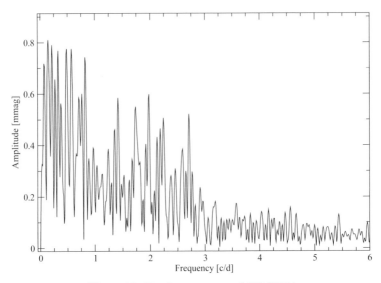

Figure 13: Fourier spectrum of HD 20884.

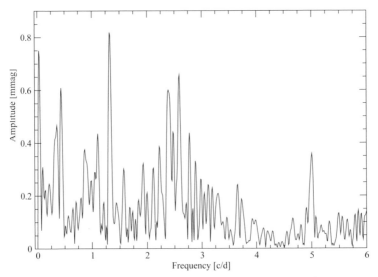

Figure 14: Fourier spectrum of HD 20790 (Guide Star g1 of the κ^1 Ceti run 2004).

The γ Dor star HD 20790

Fig. 14 shows the amplitude spectrum of this star. The significant intrinsic frequencies identified after frequency analysis and comparison to g2 and g3 are listed in Table 4. We adopt the frequency error estimation from Kallinger et al. (2008b) for the case of closely spaced frequencies, which yields an uncertainty of $0.013\,\mathrm{d}^{-1}$. The observed frequency range is typical for γ Dor pulsators. Two frequencies, f_3 and f_4, are spaced by 1 d^{-1}. Due to the clean spectral window (see Fig. 10), rather than due to aliasing, we consider these frequencies intrinsic.

In November 2004 classification spectra were taken by one of us (SR) taken at David Dunlap Observatory (DDO), using the Cassegrain spectrograph on the 1.88 m telescope. See Kallinger et al. (2008a) for the classification of all four Guide Stars of this run. The spectral classification for HD 20790 resulted in a spectral type of F4IV and a rough $v \sin i$ estimate between 80 and 130 kms^{-1}. The quality of our classification spectrum does not allow us to achieve a more precise determination of $v \sin i$. Nevertheless, we may consider this star a fast rotator.

The Strømgren calibration using the software TempLogG (Kaiser 2006) and the measurements published by Hauck & Mermilliod (1998) lead to $T_{\mathrm{eff}} \approx 7000$ K, log g ≈ 3.8 and [Fe/H] ≈ 0.04. These values place HD 20790 in the γ Dor instability strip.

The new HIPPARCOS parallax (Leeuwen 2007) of HD 20790 is 6.7 ± 0.86 mas, and the apparent magnitude given in SIMBAD is $V = 8.9$. The calculation of M_V and $L/L_{\odot}$ was performed as for HD 20884 above. The bolometric correction for early F type stars is about zero and neglected, as well as the extinction, due to the small distance. We find a luminosity range between 3.8 and 6.4 $L_{\odot}$, which is compatible with γ Dor stars.

Label	Frequency [d^{-1}]	Significance	Amplitude [mmag]
f_1	0.88	11.4	0.34
f_2	1.32	57.3	0.87
f_3	1.39	5.9	0.27
f_4	2.39	28.0	0.59
f_5	2.42	14.8	0.42
f_6	2.47	19.4	0.51
f_7	2.59	36.9	0.62
f_8	2.77	18.5	0.46

Table 4: Frequencies identified in HD 20790.

Analysis of the Guide Stars from the HR 1217 Run

The light curves show strong instrumental artifacts between JDM 1777 and 1790 not caused by the Moon, since the angular separation in this time interval was at its maximum. These artifacts introduce power at low frequencies into the Fourier spectrum. Additionally, there are a major gap of 0.9 d between JDM 1783.4 and 1784.3 as well as some minor gaps due to software crashes on-board the spacecraft. The data reduction was performed using the decorrelation method. The RMS deviation of the data sets for g1 and g2 decreased by 23.5% and only 2.2%, respectively. Although the RMS error is reduced, the point-to-point scatter is increased for g1.

Fig. 15 illustrates the effect of decorrelation for g1. The raw light curve (bottom) shows periodic spikes due to overcorrection of the background in the on-board reduction. Clipping the outliers removes these spikes, whose periodicity is the MOST orbit frequency. Due to the periodic clipping of outliers, peaks at the MOST orbit frequency and harmonics arise in the spectral window (see Fig. 16). Nevertheless, the aliasing effect does not cause significant problems to the frequency analysis, because the amplitudes in the spectral window are modest.

Fig. 17 compares three reduced light curves of stars with similar brightness (g0, g1 and g6). These stars show similar light curves and artifacts in the middle of the observations, hence they are well-suited as direct comparison stars after the data reduction. The overplotted moving medians over 300 datapoints (3.26 h) do not show periodicities for g0 and g6, while the moving median of g1 (HD 24217) displays the intrinsic pulsation including beating. The artifacts cause low-frequency power in the Fourier spectrum with similar amplitudes in the three stars, hence the instrumental signal can be identified and distinguished from stellar signal.

ID	GSC-ID	HD / BD	SpT.	V [mag]	parallax [mas]
g0	05307-00865	HD 24134	F2V	8.94	8.23 ±1.00
g1	05307-00953	HD 24217	A2m	9.23	-
g2	05307-01032	HD 24338	M2III	7.35	4.24 ±0.75
g3	05307-01312	BD -12 739	-	10.14	-
g4	05307-01021	HD 24183	F6V	7.75	13.41 ±0.61
g5	05307-01261	BD -12 738	-	9.86	-
g6	05307-00817	HD 24172	F6/F7V	9.25	-

Table 5: Information on the Guide Stars in the HR 1217 field, taken from the SIMBAD data base. The parallaxes available for three stars are taken from the new HIPPARCOS Catalogue (van Leeuwen 2007).

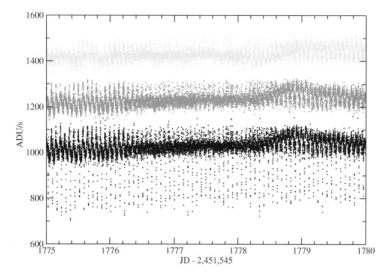

Figure 15: Effects of decorrelation on the g1 data. The light curve in the middle represents the result of a 3-σ clipping using a moving average on the raw data (bottom). The decorrelated data are displayed on top. In this plot 5 days of observation are presented for better visibility. Vertical offsets of the three plots are arbitrary.

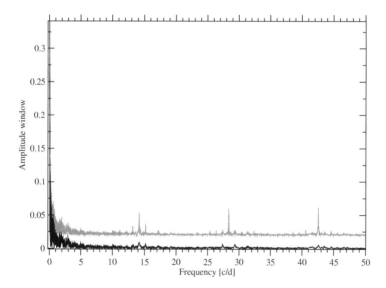

Figure 16: Comparison of the spectral windows for the raw data (black line) and the processed data (red line, offset 0.02 vertically for better visibility) for the HR 1217 field. Aliasing in the reduced Guide Star time series is modest, however present.

ID	Standard deviation [mmag]	Point-to-point scatter [mmag]
g1 raw	27.052	11.990
g1 reduced	20.691	12.520
g2 raw	66.345	15.913
g2 reduced	64.917	15.482

Table 6: Change in standard deviation and point-to-point scatter of the two variables among the HR 1217 Guide Stars. Even though the standard deviation decreases after data reduction, the point-to-point scatter increases.

Label	Frequency [d^{-1}]	Significance	Amplitude [mmag]
f_1	2.138	62.3	1.86
f_2	2.265	132	2.86
f_3	2.404	69.6	1.93
f_4	2.577	20.4	1.03

Table 7: Identified frequencies of HD 24217.

Results of the Guide Stars from the HR 1217 Run

HD 24217

The second variable star HD 24217 (g1) is an A2m star, according to SIM-BAD. The HD Catalogue classifies it as F0, whereas the Michigan Catalogue of two-dimensional spectral types for HD stars (Vol. 4) gives A2mA7-F2/3 (Houk & Smith-Moore 1998). The star shows signal in the lower frequency range around 2.26 d^{-1}, which is typical for γ Dor stars. The star g0 (HD 24134) was used as a direct comparison to identify possible instrumental effects. HD 24134 is classified as F2V and may be also a γ Dor star. However, no significant variability was detected. Fig. 18 displays the Fourier spectra of both stars over-plotted. HD 24217 shows four frequencies in the range of 2.1 to 2.6 d^{-1}. In the lower frequency range, the amplitude spectra are nearly identical. We consider the low amplitude peaks in the range of 15 to about 20 d^{-1} to be due aliasing rather than to intrinsic p-mode pulsation. The identified frequencies are given in Table 7, and the uncertainties of the frequencies are estimated to 0.009 d^{-1}.

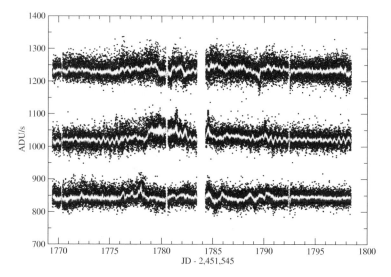

Figure 17: Reduced light curves of g0 (top), g1 (HD 24217, middle) and g6 (bottom) with a moving median over 300 datapoints (3.26 h) overplotted. These three Guide Stars are of similar brightness and show artifacts in the middle of the observing run.

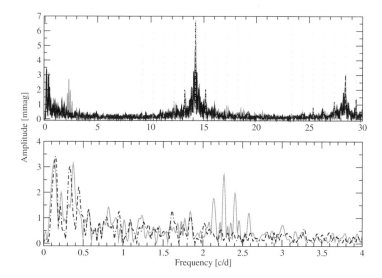

Figure 18: Fourier amplitude spectra for HD 24217 and the comparison star HD 24134. The frequency range from 0 to $30\,\mathrm{d}^{-1}$ is shown in the top panel, the range from 0 to $4\,\mathrm{d}^{-1}$ in the bottom panel. The dashed-dotted line (bottom panel) corresponds to the Fourier transform of the comparison star. The dashed-dotted vertical lines in the top panel indicate the MOST orbit frequency, its first harmonic, and $1\,\mathrm{d}^{-1}$ side lobes.

HD 24338

HD 24338 is listed in the Combined General Catalogue of Variable Stars (CGCVS, Samus et al. 2004), but no further information on its variability could be found in the literature. The light curve of HD 24338 is shown in Fig. 19. To determine the period of the long-term variability, Phase Dispersion Minimization (PDM) was applied, displayed by Fig. 20. The lowest error was obtained at a period of 17.38 d. There is a secondary minimum at 8.8 d which is due to the less pronounced secondary minimum in the light curve.

The uncertainty in the period determination is estimated to ± 0.2 d. The amplitude of the 17.4 d variation is about 40 mmag. The observed variations may well be due to large starspots and the period of 17.4 d corresponds to the rotation period of the star, based on the assumption that both deep minima in the light curves are caused by the same star spot. We also note that we only observe less than twice this variation cycle and that the second deep minimum could be caused by another spot. To derive more reliable conclusions, longer time base observations are required.

Discussion and Conclusions

A data reduction pipeline for MOST Guide Star photometry based on the decorrelation technique is presented. The pipeline consists of two IDL programs, one for data extraction and one for data reduction. Since MOST observes at least four Guide Stars in one field simultaneously, common instrumental signal can be identified and mitigated, in some cases even by a factor of 10. If appropriate comparison stars are available, three-star photometry, which is common for ground based observations, can be applied. However, in many cases comparison stars are much too faint or are affected quite differently by stray light. The decorrelation method can deal with these conditions in a much better way.

The number of decorrelation steps is crucial for the method discussed here. If many non-variable Guide Stars are available, many decorrelation steps can be performed, which results in good final light curves shown in Fig. 3. To the contrary, if only a few reference stars are available, the results may not be significantly better than the raw light curves after outlier clipping, especially if the comparison stars are rather faint, like in the presented case of κ^1 Ceti 2004 Guide Stars. However, a considerable reduction of the time series' variance can be achieved, if the variable and the comparison stars are similar in brightness.

Data smoothing techniques, which are applied to Guide Stars individually, have the disadvantage that the information provided by other stars is ignored. Bandpass filters, for example periodic filter functions, may strongly decrease the stellar signal at frequencies close to the instrumental frequencies, which makes the filter techniques unsuited for data variables with low-frequency signal like γ Dor stars or signal in the vicinity of the MOST artifacts like δ Scuti stars.

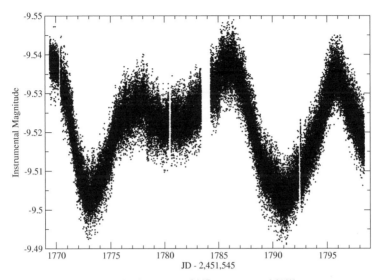

Figure 19: Light curve of HD 24338, an M2III star.

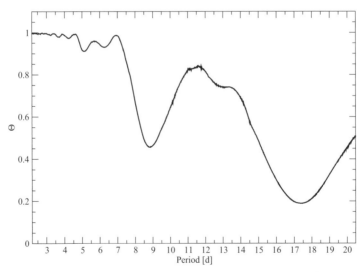

Figure 20: PDM method applied to the light curve of HD 24338. The minimum of
the error is at about 17.38 days, and a secondary minimum is present at 8.83 d.

The MOST Guide Star photometry presented here contains four variables: (1) one pulsating K giant (HD 20884) and (2) one M giant most probably showing large starspots. Assuming that the first and the second deep minimum in the light curve of the M giant is due to the same spot, the rotation period of this star would be about 17.4 d. Additionally, we report on two new γ Dor star discoveries, (3) HD 20790 and (4) HD 24217. Our classification spectrum and the published Strømgren colors justify the γ Dor classification for HD 20790, while no published Strømgren colors and no classification spectrum for HD 24217 are available. Moreover the spectral classification in the literature, especially the classification in the Michigan Catalogue A2mA7-F2/3 indicates chemical peculiarities of this star. A few high-resolution spectra for abundance analysis could help to clarify this situation and the possible binarity.

Acknowledgments. MH and WW are supported by the Austrian Fonds zur Förderung der wissenschaftlichen Forschung (FWF, project *The Core of the HR diagram*, P 17580-N02) and the Bundesministerium für Verkehr, Innovation und Technologie (BM.VIT) via the Austrian Agentur für Luft- und Raumfahrt (FFG-ALR). JMM, SR, DBG, and AFJM received research support from NSERC (Natural Sciences & Engineering Research Council) Canada. RK was partly funded by the Canadian Space Agency. DS is supported by the National Science Foundation.

References

ESA 1997, *The Hipparcos and Tycho Catalogues,* ESA-SP 1200

Hauck, B., Mermillod, M. 1998, A&AS, 129, 431

Houk, & Smith-Moore 1998, Michigan Catalogue of HD stars Vol. 4

Huber, D., & Reegen, P. 2008, CoAst, 152, 77

Kaiser, A. 2006, ASPC, 349, 257

Kallinger, T., Guenther, D. B., Weiss, W. W., et al. 2008a, CoAst, 153, 84

Kallinger, T., Reegen, P., & Weiss, W. W. 2008b, A&A, 481, 571

Matthews, J. M., Kuschnig, R., Guenther, D. B., et al. 2004, Nature, 430, 51

Reegen, P., Kallinger, T., Frast, D., et al. 2006, MNRAS, 367, 1417

Reegen, P. 2007, A&A, 467, 1353

Rowe, J. F., Matthews, J. M., Kuschnig, R., et al. 2006, MmSAI, 77, 282

Rucinski, S. M., Walker, G. A. H., Matthews, J. M., et al. 2004, PASP, 116, 1093

Samus, N. N., Durlevich, O. V., et al. 2004, Combined General Catalogue of Variable Stars

van Leeuwen, F. 2007, A&A, 474, 653

Walker, G. A. H., Matthews, J. M., Kuschnig, R., et al. 2003, PSAP, 115, 1023

Comm. in Asteroseismology
Vol. 156, 2008

First asteroseismic results from CoRoT

E. Michel[1], A. Baglin[1], W.W. Weiss[2], M. Auvergne[1], C. Catala[1], C. Aerts[3],
T. Appourchaux[4], C. Barban[1], F. Baudin[4], M. Briquet[3], F. Carrier[3],
P. Degroote[3], J. De Ridder[3], R.A. Garcia[5], R. Garrido[6], J. Gutiérrez-Soto[1,7],
T. Kallinger[2], L. Lefevre[1], C. Neiner[7], E. Poretti[8], R. Samadi[1], L. Sarro[9],
G. Alecian[10], L. Andrade[11], J. Ballot[12], O. Benomar[4], G. Berthomieu[13],
P. Boumier[4], S. Charpinet[14], B. de Batz[7], S. Deheuvels[1], M.-A. Dupret[1],
M. Emilio[15], J. Fabregat[16], W. Facanha[11], M. Floquet[7], Y. Frémat[17],
M. Fridlund[18], M.-J. Goupil[1], A. Grotsch-Noels[19], G. Handler[2], A.-L. Huat[7],
A.-M. Hubert[7], E. Janot-Pacheco[11], H. Kjeldsen[20], Y. Lebreton[7], B. Leroy[1],
C. Martayan[7,17], P. Mathias[21], A. Miglio[19], J. Montalban[19],
M.J.P.F.G. Monteiro[22], B. Mosser[1], J. Provost[13], C. Regulo[23], J. Renan de
Medeiros[24], I. Ribas[25], T. Roca Cortés[23], I. Roxburgh[26,1], J. Suso[16],
A. Thoul[19], T. Toutain[27], D. Tiphene[1], S. Turck-Chieze[5], S. Vauclair[14],
G. Vauclair[14], K. Zwintz[2]

(Affiliations can be found after the references)

Abstract

About one year after the end of the first observational run and six months after the first CoRoT data delivery, we comment the data exploitation progress for different types of stars. We consider first results to illustrate how these data of unprecedented quality shed a new light on the field of stellar seismology.

Individual Objects: HD 50747, HD 49933, HD 50890, HD 50170, HD 51106, HD 181420, HD 181906, HD 180642, HD 181231, HD 175869, HD 181907

The present CoRoT harvest

The detailed description of the instrument and programme has been presented in several places (e.g. Baglin et al. 2006, Auvergne et al. 2006). Detailed measurements of the in-flight performances are now available (Auvergne et al. 2008). We simply remind you here of a few aspects necessary to introduce the results.

The CoRoT focal plane is divided in two areas of about 4 square degrees on the sky each. One of them (the seismo field) is mostly optimized for the seismology programme (Michel et al. 2006). Ten objects with $5.4 < m_V < 9.5$ can be observed simultaneously with a 1 second-sampling rate and a noise limited by photon noise to $4\ 10^{-4} < \sigma < 3\ 10^{-3}$ per 1 s measurement.

The other field (the exofield) is mostly optimized for the search of exoplanets. About 12,000 objects with $11 < m_V < 15$ can be observed simultaneously, with a 512 s sampling rate and a standard deviation from $7\ 10^{-4}$ to $2\ 10^{-3}$ per 512 s measurement. In this field, for the brightest stars ($m_V < 14.5$), a prism allows to obtain three-colour information. For a limited number ($\sim$500), it is possible to select a 32 s sampling rate.

Altogether, during the 3-years nominal period foreseen for the mission, more than 100 stars will be observed in the Seismo field and more than 100 000 in the exofield!

As described in Michel et al. (2006b), the observational programme is composed of long runs (up to 150 days) and short runs (20-30 days). At the time of the Wrocław conference, 5 runs have been completed successfully: an initial run of 60 days, two long runs and two short runs. The satellite is presently in a long run which will stop by mid-october.

We thus have completed the observations of 50 objects in the seismo field (in fact 49, HD 49933 having been observed twice): 5 solar-like pulsator candidates; 13 B stars, including 1 known Beta Cephei star, 5 Be stars, 2 eclipsing binaries; 7 F and G giant stars ; 17 A stars, including 4 known pulsators, 2 Am stars, 3 Ap stars, 2 eclipsing binaries; 7 early F stars including 2 known Gamma Dor.

A few examples of the light curves are given in Fig. 1 and Fig. 2. They illustrate the quality of the data, revealing in the light curves variability below the 10^{-3} level. They also illustrate the variety of variability behaviours encountered.

Solar-like pulsators

The search and characterization of solar-like oscillations in other main sequence stars is one of the highlights and one of the greatest challenges of the CoRoT mission. The main objective is to bring measurements of the modes parameters (frequencies ,amplitudes, mode profiles,...) for objects scanning the Main Sequence in the range of temperature where such oscillations are expected.

An example of light curves obtained for one of these candidates (HD 49933) is given in Fig. 2. In this light curve, the standard deviation of the individual measurements (every 32 s here) is of the order of 10^{-4} (about 100 times below the standard values from ground-based photometry). The light curve already reveals variability of a few 10^{-4} which is attributed to activity and interpreted in terms of spots. The solar-like oscillations do not appear in the light curve but are clearly seen in the Fourier spectra (Fig. 3) for three objects of various magnitudes.

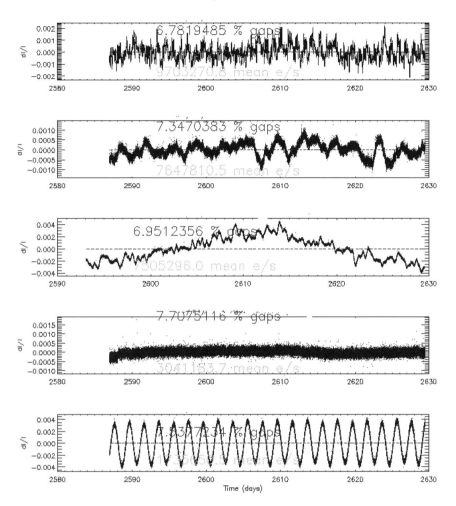

Figure 1: Forty-days-portions of the light curves from 5 stars (top to bottom: the A star HD 50747, the F star HD 49933, the giant G star HD 50890, the giant F star HD 50170, the A star HD 51106)

As shown by Michel et al. (2008) and illustrated in Fig. 3, it is possible to identify three components in the spectra of each of these three stars: (i) a white component compatible with the photon noise; (ii) a component increasing toward low frequencies characterizing the stellar granulation and (iii) the oscillations component with its characteristic comb-like pattern. Michel et al. (2008) give a comparison of these measurements with those observed in the Sun and with theoretical values.

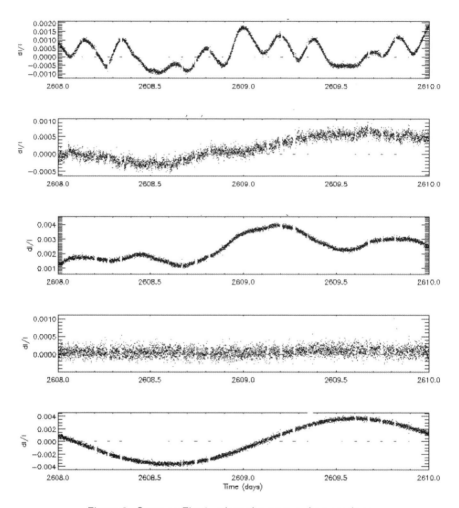

Figure 2: Same as Fig. 1 enlarged over two-days-portions

The activity of data analysis intended to provide for each star individual mode parameters (frequencies, amplitudes, profiles,...) is held within the CoRoT/SWG/DAT (resp. T. Appourchaux, see Appourchaux et al. (2006)). An example of analysis can be found in Appourchaux et al. (2008), where about 40 modes have been characterized for HD 49933, including eigenfrequencies determined with precision of the order of a few 10^{-7} Hz to a few 10^{-6} Hz (Fig. 4). Further analyses with alternative methods are under way as illustrated by Garcia et al. (2008).

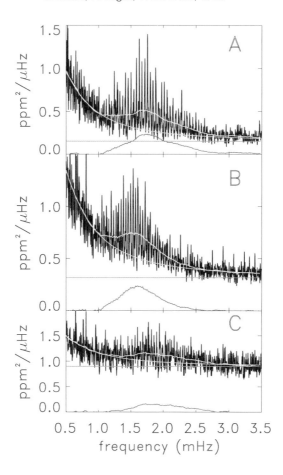

Figure 3: Power spectra of three solar-like pulsators observed with CoRoT: HD 49933 ($m_V = 5.7$), HD 181420 ($m_V = 6.7$), and HD 181906 ($m_V = 7.6$). Each spectrum is decomposed in three contributions commented in the text (after Michel et al. 2008)

B stars

As in most of the classical (opacity driven) pulsators, the gain in resolution and noise level is expected to bring a new insight on the pulsational behaviour of the stars and bring precise measurements for seismic studies. In addition, thanks to their continuity, the CoRoT data also allow to tackle oscillations with periods of the order of the day or below, which are very common in Slowly Pulsating B stars and Beta Cephei pulsators.

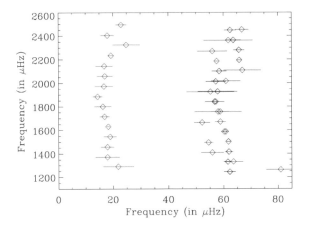

Figure 4: Echelle diagram showing the frequencies of the modes and their 3 sigma-error bars obtained for HD 49933 (after Appourchaux et al. 2008)

The data analysis of these targets is organized within the CoRoT B stars group (Resp. C. Aerts). As an example, the Fourier spectrum of the known Beta Ceph HD 180642 reveals a dominant radial mode with its harmonics but also low amplitude nonradial modes (Fig. 5).

Among the latter, the highest ones have also been measured in spectroscopy (with FEROS, in the framework of the ESO Large Programme, led by E. Poretti) which will bring helpful complementary information about their identification. More details on this analysis will be given in Briquet et al. (2008).

Be stars

The analysis lead in the CoRoT/Be stars group (resp. C. Neiner) allowed to reveal oscillations in a B5IVe star (Fig. 6 and Neiner et al. 2008). Only two examples of such oscillation detection in late Be stars existed so far from MOST photometry (Walker et al. 2005, Saio et al. 2007).

For another object, HD 175869, peaks have been detected with amplitudes down to a few 10^{-6}. A preliminary study suggests that these peaks are associated with oscillations and rotation (Gutierrez-Soto 2008).

A and early F stars

A and early F type stars feature several known classes of variables among which δ Scuti stars and γ Doradus stars are the most famous. For the former, the objects observed from the ground show up to several tens of modes, but always less than expected from theoretical models. This raised the question of the

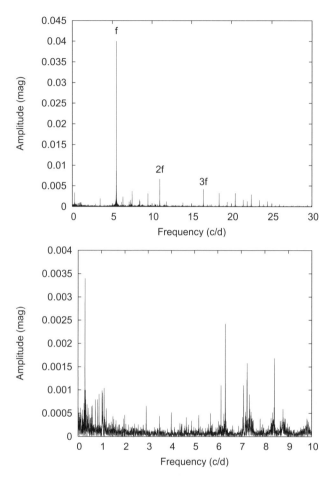

Figure 5: top: Power spectrum of the Beta Cephei HD 180642; bottom: idem after removing the radial mode and its harmonics (after Briquet et al. 2008)

possible existence of selection rules favorising given modes versus others. However, in the existing observations, the detection is always limited in amplitude to a few 10^{-4} by the noise level. By lowering the noise level by a factor 100, the CoRoT data are expected to bring a new insight on this question.

The light curve of HD 174936 reveals variations at the 10^{-3} level in the envelope, with beating phenomenon (Fig. 7). The associated Fourier spectrum shown in Fig 8 reveals a very rich spectrum. After fitting and subtracting peaks above an illustrative $4\ 10^{-4}$ level, the residual spectrum (Fig 9) reveals

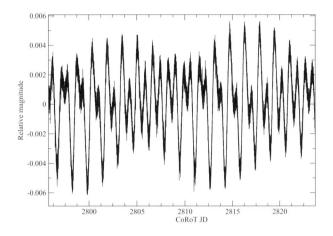

Figure 6: light curve of the B5IVe star HD 181231 (after Neiner et al. 2008)

a high number of remaining peaks, much higher than the noise level as estimated from the Fourier spectrum of a star of same brightness showing no signal in this frequency range. The interpretation of these results is led within the CoRoT/A stars group (resp. R. Garrido and E. Michel).

Giant F and G stars

For these stars, solar-like oscillations are expected with amplitudes significantly larger than in the main sequence solar-like pulsators and with frequencies in the 10-100 μHz domain. Here again, the CoRoT data allow to reveal these oscillations (e.g. Fig. 10).

Some of these objects show more surprising behaviour. One of them (HD 50170) shows no apparent variability in its light curve (Fig. 2). Its spectrum reveals two types of peaks: narrow peaks which would be consistent with long-lived modes forming an l=1 multiplet (Fig. 11) and at a lower amplitude, modes with shorter lieftimes, as suggested by the autocorrelation function (Fig. 11) A paper by Baudin et al., which is in preparation, will discuss this in more detail. This object is being investigated in order to determine whether we are dealing with a new type of hybrid pulsator or these pulsations have to be attributed to separate components of a multiple star.

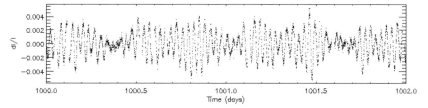

Figure 7: light curve (part of) of the δ Scuti star HD 174936.

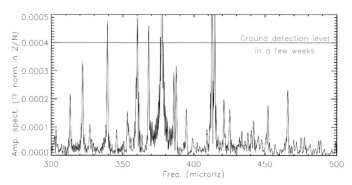

Figure 8: Amplitude spectrum of the δ Scuti star HD 174936.

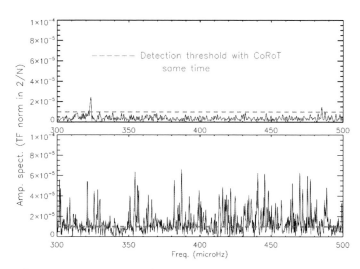

Figure 9: Bottom: same as Fig. 8 after substraction of the highest peaks; top: amplitude spectrum for a comparison star of similar magnitude showing no oscillations in this frequency range

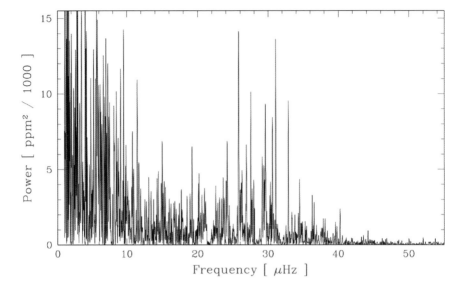

Figure 10: Power spectrum of the giant star HD 181907 showing oscillations in the range 20-40 μHz

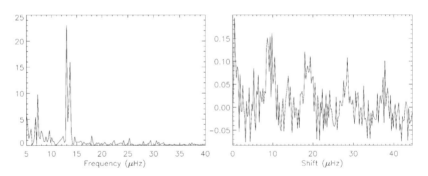

Figure 11: left: Power spectrum of the giant star HD 50170; right: Autocorelation of the power spectrum

In the exofield

About 60 000 objects have been observed in the exofield so far. While they are actively searched for occultation due to planet transits, these data also constitute a goldmine for seismology. The activity of seismology on these objects is organized within the Additional Programme Working Group (APWG, led by W. W. Weiss).

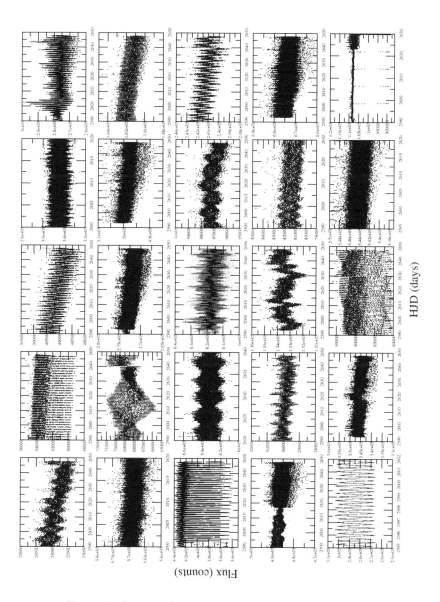

Figure 12: Sample of light curves obtained in the exofield

While some of the light curves obtained in the exofield show variations which can easily be recognized as characteristics of specific type of pulsators, others show much more puzzling behaviour (Fig. 12).

Due to the large number of objects considered and to the fact that very few of them have well-defined global parameters (due to their faintness), an activity of automatic classification has been undertaken in two groups (at Leuven and Madrid). These processes are based on self-learning algorithms, and the results obtained so far come from the first loop and are expected to evolve significantly. However, it seems interesting to note that among the few thousands of variables already found, the different known classes of pulsators are represented as well as apparently new classes (Degroote et al. 2008).

The red giants bring a striking example of the potential of the exofield for stellar seismology (led by J. de Ridder). As shown in Kallinger et al. (2008) (see also Hekker et al. 2008), solar-like oscillations are observed in a significant

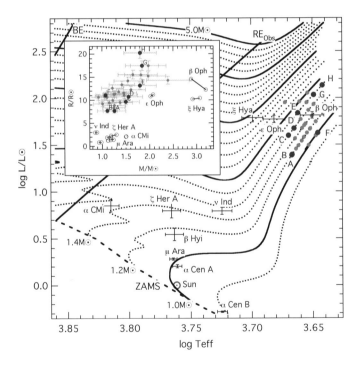

Figure 13: Location of several giant pulsators in an HR diagram and in a mass-radius diagram derived from observed oscillations (after Kallinger et al. 2008)

number of red giants revealing trends linked with their global parameters. With these data, Kallinger et al. (2008) have been able to estimate mass and radii for those objects (Fig. 13).

Conclusions

The CoRoT data fulfil expectations in terms of noise level, duration of the runs and continuity of the observations. The analysis of the light curves allows to explore stellar variability at an unprecedented level of precision and for an unprecedented range of time scales. The first interpretation studies confirm the great help we can expect from these data to improve our understanding of stellar structure and evolution. These results also confirm space photometry as an efficient component of the strategy to develop stellar seismology.

Acknowledgments. The CoRoT (Convection Rotation and planetary Transits) space mission, launched on December 2006, was developed and is operated by CNES, with participation of the Science Program of ESA, ESA's RSSD, Austria, Belgium, Brazil, Germany and Spain. The Austrian Science Fonds has supported WWW and TK (P17890), and KZ (T335-N16). MB is Postdoctoral Fellow of the Fund for Scientific Research, Flandres.

References

Appourchaux, T., Berthomieu, G., Michel, E., et al. 2006, The CoRoT Mission, Eds. Fridlund M., Baglin A., Lochard J. and L. Conroy, ESA SP-1306, 377

Appourchaux, T., Michel, E., Auvergne, M., et al. 2008, A&A, 488, 705

Auvergne, M. 2006, The CoRoT Mission, Eds. Fridlund M., Baglin A., Lochard J. and L. Conroy, ESA SP-1306, 283

Auvergne, M., Boisnard, L., Lam-Trong, T., et al. 2008, A&A, in press

Baglin, A., Auvergne, M., Barge, P., et al. 2006, The CoRoT Mission, Eds. Fridlund M., Baglin A., Lochard J. and L. Conroy, ESA SP-1306, 33

Briquet, M., Uytterhoeven, K., Aerts, C., et al. 2008, CoAst, 158, in press

Degroote, P., Miglio, A., Debosscher, J., et al. 2008, CoAst, 158, in press

Garcia, R., Appourchaux, T., Baglin, A., et al. 2008, CoAst, 157, in press

Gutierrez-Soto, J., Neiner, C., Hubert, A-M., et al. 2008, CoAst, 157, in press

Hekker, S., Barban, C., Hattzes, A., et al. 2008, CoAst, 157, in press

Kallinger, T., Weiss, W. W., Barban, C., et al. 2008, A&A, in press

Michel, E., Baglin, A., Auvergne, M., et al. 2006, The CoRoT Mission, Eds. Fridlund M., Baglin A., Lochard J. and L. Conroy, ESA SP-1306, 39

Michel, E., Deleuil, M., & Baglin, A. 2006b, The CoRoT Mission, Eds. Fridlund M., Baglin A., Lochard J. and L. Conroy, ESA SP-1306, 473

Michel, E., Baglin, A., Auvergne, M., et al. 2008, Science, 322, 558

Neiner, C., Gutierrez-Soto, J, Fremat, Y., et al. 2008, CoAst, 157, in press

Saio, H., Cameron, C., Kuschnig, R., et al. 2007, ApJ, 654, 544

Walker, G. A. H., Kuschnig, R., Matthews, J. M., et al. 2005, A&A, 635, 77

Affiliations of the authors

[1] LESIA-Observatoire de Paris-CNRS (UMR 8109)-Univ. Paris 6- Univ. Paris 7,
    pl. J. Janssen, F-92195 Meudon, France

[2] Institute for Astronomy Univ. of Vienna,
    Tuerkenschanzstrasse 17, A-1180 Vienna, Austria

[3] Instituut voor Sterrenkunde, Katholieke Univ. Leuven,
    Celestijnenlaan 200 D, B-3001 Leuven, Belgium

[4] Institut d'Astrophysique Spatial, Univ. Paris 11 - CNRS (UMR 8617),
    F-91405 Orsay, France

[5] Labo. AIM, CEA/DSM-CNRS-Univ. Paris 7; CEA, IRFU, SAp,
    Centre Saclay F-91191 Gif-sur-Yvette, France

[6] Instituto de Astrofisica de Andalousia -CSIC,
    Camino Bajo de Huetor, 50, E-18008 Granada, Spain

[7] GEPI-Observatoire de Paris-CNRS (UMR 8111)- Univ. Paris 7,
    pl. J. Janssen, F-92195 Meudon, France

[8] INAF-Osservatorio Astron. di Brera, Via E. Bianchi 46, I-23807 Merate (LC)
    Italy

[9] LAEFF, Apt. 78, E-28691 Villanueva de la Cañada Madrid, Spain

[10] LUTH-Observatoire de Paris-CNRS-Univ. Paris 7,
    pl. J. Janssen, F-92195 Meudon, France

[11] Universidade de Sao Paulo, Rua do Matão, 1226 São Paulo-SP Brazil

[12] Max Planck Inst. für Astrophysik,
    Karl-Schwarzschild-Str. 1, Postfach 1317, 85741 Garching, Germany

[13] Laboratoire Cassiopée, Observatoire de la Côte d'Azur-Univ. Nice-Sophia
    Antipolis-CNRS (UMR 6202), BP 4229, F-06304 Nice, France

[14] Laboratoire d'Astrophysique de Toulouse-Tarbes- Univ. de Toulouse-CNRS,
    14, av. E. Belin, F-31400 Toulouse, France

[15] Universidade Estadual de Ponta Grossa,
    Praca Santos Andrade, 1, Ponta Grossa- Paraná- Brazil

[16] Observatori Astronómic de la Universitat de Valéncia,
    Poligon La Coma, 46980 Paterna Valéncia

[17] Royal Observatory of Belgium, 3 av. circulaire, 1180 Brussel, Belgium

[18] Astrophysics Mission Division, RSSD ESA, ASTEC, SCI-SA P.O. Box 2999, Keplerlaan 1 NL-2200AG, Noordwijk, The Netherlands

[19] Department of Astrophysics and Geophysics-FNRS, Liège Univ., 17 allée du 6 août, 4000 Liège, Belgium

[20] Danish AsteroSeismology Centre (DASC), Department of Physics and Astronomy, Univ. of AArhus, 8000 Aarhus C, Denmark

[21] Laboratoire Fizeau, Observatoire de la Côte d'Azur-CNRS (UMR 6525)-Univ. Nice Sophia Antipolis, campus Valrose, F-06108 Nice, France

[22] DMA/FCUP & Centro de Astrofísica da Univ. do Porto, Ruas das Estrelas, 4150-762, Porto, Portugal

[23] Instituto de Astrofísica de Canarias - Departamento de Astrofísica, Univ. de La Laguna, Tenerife, Spain

[24] Departamento de Física, Univ. Federal do Rio Grande do Norte, 59072-970, Natal RN, Brasil

[25] Institut de Ciencies de l'Espai (CSIC-IEEC) Campus UAB Facultat de Ciències, Torre C5-parell, 2a pl 08193 Bellaterra, Spain

[26] Astronomy Unit, Queen Mary, Univ. of London, Mile End Road, London E1 4NS, UK

[27] School of Physics and Astronomy, Univ. of Birmingham, Edgbaston B15 2TT, UK

HELAS
News

Comm. in Asteroseismology
Vol. 156, 2008

Announcement of HELAS III

C. Aerts[1] for the HELAS Board[2]

[1] Instituut voor Sterrenkunde, Celestijnenlaan 200D, B-3001 Leuven, Belgium

[2] The HELAS Project Office, Kiepenheuer-Institut für Sonnenphysik, Schöneckstr. 6, D-79104 Freiburg, Germany; http://www.helas-eu.org

Abstract

We announce the First CoRoT International Symposium, which is a joint organisation between CNES, HELAS and Observatoire de Paris.

The CoRoT space mission (http://corot.oamp.fr/) was launched successfully on 27 December 2006. It has been developed and is operated by the French space agency CNES, with contributions from Austria, Belgium, Brasil, ESA, Germany and Spain. While the first CoRoT data have only been released to the consortium members with proprietory data rights so far, it is already clear from the first results that CoRoT will bring a revolution to the fields of exoplanet research and asteroseismology the coming years.

One of the network activities of the HELAS Coordination Action is to organise four large international symposia in Europe. In view of the importance of the CoRoT space mission, the HELAS Board has chosen to collaborate with the CoRoT consortium for one of those four meetings, and to dedicate its third symposium to the first public release of CoRoT results and data. We are thus happy to announce the *First CoRoT International Symposium* as a joint organisation between CNES, HELAS and Observatoire de Paris. It will take place from 2 to 5 February 2009 in Paris at the Maison Internationale. Regular updates of information will be available from: http://www.symposiumcorot2009.fr/

The HELAS contribution to the organisation will be the funding of the proceedings, which will be a special volume of the journal *Astronomy & Astrophysics*, as well as the provision of grants to PhD students and young postdocs. Information on these items will be made available at the above conference website.

Comm. in Asteroseismology
Vol. 156, 2008

Conference Review of the
38th LIAC / HELAS-ESTA / BAG
Liége, Belgium, July 7-11 2008

Evolution and Pulsation of
Massive Stars on the Main Sequence
and Close to it

A. Grötsch-Noels,[1] J. Montalban,[1] and A. Miglio[1]

[1] Institut d'Astrophysique et de Géophysique
Université de Liège, Alle du 6 Août 17 - B 4000 Liège - Belgique

Website : http ://www.ago.ulg.ac.be/APub/Colloques/Liac38/

Objectives and scientific rationale

The main idea of this colloquium was first, to close in on the problems raised by "nonstandard" physics, second, to focus on the effects of these often missing physical processes on stellar evolution and third, to analyze what asteroseismology could do to shed a new light on these processes and their modeling. The targets are massive stars (O, B, WR) for which new and exciting results are now coming from asteroseismic interpretations of observed modes.

The main objectives were threefold. In a first part, physical processes involved in "nonstandard" modeling such as semiconvection, overshooting and convective penetration, rotation, diffusion,... were analyzed in detail. A second part focused on the problems related to the effect of such "nonstandard" processes on stellar modeling. The third part introduced asteroseismology as a probe of the internal structure of stars, in particular in the frame of present and future asteroseismology missions, from space or ground-based, such as CoRoT, KEPLER, PLATO, SIAMOIS, SONG,...

In what follows, we present a flavor of the topics discussed in seven sessions covering these three parts.

Part 1 : Internal structure of massive stars

Session 1 - Physics and uncertainties in massive stars on the main sequence and close to it

The longstanding problems in hot star's internal modeling such as convection and semiconvection, rotation, overshooting, diffusion and radiative forces,... are analyzed in order to give the audience a good theoretical view on each problem without entering the intricacy of stellar modeling.

Session 2 - Physics and uncertainties and their effects on the internal structure

Each of these aspects influences the models. To what extent? This second session is a critical analysis of standard modeling versus nonstandard modeling. What do we expect?

Part 2 : Outer layers of massive stars

Session 3 - Atmosphere, mass loss and stellar winds

This session deals with the latest results about the physics and modeling of the expanding outer layers of massive stars. The question : To what extent the observations of the expanding atmosphere transposable to the photosphere? In particular, are the observed periods global oscillations of the star or do they find their origin in the atmosphere itself?

Part 3 : Asteroseismology of massive stars

Session 4 - Observed frequencies in pulsating massive stars

This session is devoted to observational asteroseismology of hot stars, gathering of ground-based and space data and their interpretation.

Session 5 - What can asteroseismology do to solve the problems?

What are the asteroseismic signatures of all these physical processes? What can we learn from the driving mechanisms? What about the limits of the convective cores? What about the treatment of semiconvection? What about rotation? Is there a connection between some violent instabilities and mass loss?

Session 6 - What about real stars?

This session is essentially focused on the theoretical calibration of ground-based, MOST and CoRoT targets.

Special session : Future asteroseismic missions

The advancement of the future asteroseismic missions (KEPLER, PLATO, SIAMOIS, SONG) are presented by experts deeply involved in their conception and preparation.

Comm. in Asteroseismology
Vol. 156, 2008

HELAS Local Helioseismology Activities

H. Schunker, and L. Gizon

Max-Planck-Institut für Sonnensystemforschung,
Max-Planck-Strasse 2, 37191, Katlenburg-Lindau, Germany

Abstract

The main goals of the HELAS local helioseismology network activity are to con-solidate this field of research in Europe, to organise scientific workshops, and to facilitate the distribution of observations and data analysis software. Most of this is currently accomplished via a dedicated website at http://www.mps. mpg.de/projects/seismo/NA4/. In this paper we list the outreach mate-rial, observational data, analysis tools and modelling tools currently available from the website and describe the focus of the scientific workshops and their proceedings.

Introduction

Local helioseismology studies the three-dimensional structure of sunspots and active regions, local mass flows and enables us to identify magnetic activity on the far-side of the Sun. Many properties of the solar interior that may be probed with local helioseismology have yet to be revealed. At the moment abundant high-quality helioseismic data exist from the Global Oscillation Network Group (GONG) and the Michelson Doppler Imager (MDI) instrument to work with. In the near future, the Solar Dynamics Observatory (SDO) will be launched with the Helioseismic and Magnetic Imager (HMI) on board, which will collect full-disk, high-resolution data. To support European local helioseismology and to exploit the large amounts of data efficiently, it is helpful to select and make available useful observations and analysis tools.

The European Helio- and Asteroseismology Network (HELAS) was created to co-ordinate the exchange of knowledge, data and software tools amongst researchers. There are four main network activities. One of these network activities (NA4) carries out the objectives of HELAS specifically for local he-lioseismology. It is implemented via a dedicated website and through meetings and workshops.

In the following section, we discuss the website which holds about 1 TB of data and provides a platform for the exchange of observations, analysis tools, solar models and general information about local helioseismology. The work-shops and meetings are described towards the end of this paper. Finally, we list the publications pertaining specifically to HELAS local helioseismology.

The HELAS local helioseismology website

The address of the HELAS local helioseismology website is http://www.mps. mpg.de/projects/seismo/NA4/ and is hosted by the Max Planck Institute for Solar System Research (MPS), in Katlenburg-Lindau, Germany. It boasts straightforward access to specially selected useful and interesting helioseismic data sets, outreach material, tools to analyse the observations and modelling codes suitable for local helioseismology. Since September 2007, there have been 380 absolute unique visitors from 64 countries. The website is regularly updated with new material, when it becomes available. The system is backed up regularly, ensuring ongoing availability of the data from the website. This section goes through the main parts of the website.

Documentation and outreach material

The website is aimed at both students and professional scientists in the field. We provide links to educational material comprising media coverage, technical reports and scientific papers at http://www.mps.mpg.de/projects/seismo/ NA4/helasNA4_General.html. The media coverage includes science maga-zines and news reports in a variety of languages. The technical information incorporates review papers, a list of appropriate text books, lecture notes and PhD theses. We also link to the homepages of academic groups active in the field. The outreach section should provide a scientist or a student with enough information to understand the basics of helioseismology and its importance.

Observations

One of the purposes of HELAS local helioseismology is to select useful observa-tional data sets and make them available. In this section, the selected data sets that are currently available at http://www.mps.mpg.de/projects/seismo/ NA4/DATA/data_access.html are listed. An earlier description can be found in Schunker et al. (2008). The observations provided on the website are predom-inantly Dopplergrams, magnetograms, intensity continuum images and vector

European Helio- and Asteroseismology Network (HELAS)

Local Helioseismology Network Activity

Welcome

The main goals of the HELAS Local Helioseismology Network Activity are to consolidate this field of research in Europe, to organize scientific workshops, and to facilitate the distribution of observations and data analysis software.

What is local helioseismology & why is it important? Solar oscillations enable helioseismologists to see inside the Sun, just as geophysicists can probe the internal structure of the Earth using records of seismic activity. Local helioseismology is being developed to make three dimensional images of the solar interior. Detailed maps of the upper convection zone provide new insights into the structure, evolution and organization of active regions and convective flows. Local helioseismology is also used to construct maps of active regions on the other side (where we cannot observe) of the Sun. Click here for an extended description.

Figure 1: Homepage of the HELAS local helioseismology website which acts as a platform for the exchange of knowledge, data, analysis tools and modelling tools at http://www.mps.mpg.de/projects/seismo/NA4/.

magnetograms from the Solar and Heliospheric Observeratory's (SOHO) MDI, the GONG and the Magneto-Optical filter at Two Heights (MOTH) instruments in Flexible Image Transport System (FITS) format. The observations cover entire Carrington rotations as well as selected active and quiet Sun regions. The observational data sets are listed in a quick look format including a sample graphic and a description of the data, as shown in the data table in Figure 2. We provide a dedicated web page for each data set with more detailed information relevant to the particular observations.

The most complete data set available from the HELAS local helioseismology website are the observations of AR9787. The data consists of nine full days of MDI Doppler velocity and magnetic field observations of the large, isolated sunspot in AR9787 with a cadence of one minute. One intensity continuum image exists for each six hours. Solar Flare Telescope (SFT) vector magnetograms are also available in this data set. This data set was analysed extensively at the second HELAS local helioseismology workshop in January 2008 and the results are published in Gizon et al. (2008).

The rest of the data sets available from the HELAS local helioseismology website are now listed: (1) Doppler velocity observations from the GONG mapped tiles throughout Carrington Rotations 1988 and 2024; (2) MDI high resolution, thirty-five hour Doppler velocity observations of AR8555 and a por-

Name	Quick Look	Description	Type	Time	AR	Size/Res.	Extra Information
AR9787	•	A relatively large, isolated sunspot.	MDI, I(fd), V(fd), M(fd) inc. Vec. Mag., SFT Vec. Mag.	20.01.2002-28.01.2002 (24x9 hrs)	Solar Monitor images of AR9787	512x512x1440 pix (9) 0.12 deg/pix	Extra info
CR1988		Tiles at various lat/lon 30 March 2002 - 26 April 2002	GONG, V	28.03.2002-27.04.2002	Carrington Rotation map	128x128 pix 0.125 deg/pix	Extra info
CR2024		Tiles at various lat/lon 5 December 2004 - 2 January 2005	GONG,V	04.12.2004-04.01.2005	Carrington Rotation map	128x128 pix 0.125 deg/pix	Extra info
AR8558		Region with a sunspot and plenty of quiet Sun	MDI, V(hr), M(fd), I(fd)	02.06.1999 (35 hrs)	Solar Monitor images of AR8558	512 x 256 x 2100 pix 0.068 deg/pix	Extra info
AR9236		A large complex sunspot in high-resolution.	MDI, V(hr), SFT Vec. Mag.	22.11.2000-24.11.2000 (56 hrs)	Solar Monitor images of AR9236	512 x 512 x 3360 0.068 deg/pix	Extra info
AR8403, AR8402		Three sunspots in AR 8403, 8402 located close together.	MDI, I(hr), V(hr), M(hr)	06.12.1998 (8 hrs)	Solar Monitor images of the three sunspots	512 x 256 x 480 pix 0.068 deg/pix	Extra info
Quiet Sun		Quiet Sun (with a hint of AR8116 in top left).	MDI, I(hr), V(hr), M(hr)	07.12.1997 (4 hrs)	Active region map	900 x 500 x 240 pix 0.068 deg/pix	Extra info
South Pole Data		South Pole observations. Different heights of the atmosphere, K, Na, Ni.	MOTH, K, Na, MDI Ni	20.01.2003 (~17 hrs)	Active region map	123 x 123 x 6416 pix 0.3089 deg/pix	Extra info

Figure 2:   Quick look data table of some of the available data sets
http://www.mps.mpg.de/projects/seismo/NA4/DATA/data_access.html.

tion of quiet Sun as well as full-disk intensity and magnetograms; (3) Observations of a large, complex sunspot, AR9236, observed with MDI high-resolution
for 56 hours is accompanied by SFT vector magnetograms; (4) MDI high-
resolution Doppler, magnetic and intensity continuum observations of three
sunspots in close proximity observed over eight hours; (5) Four hours of MDI
high-resolution Doppler velocity observations of quiet Sun; (6) Multiple spectral line observations from the MOTH instrument provide unique simultaneous
multi-height observations of the same spatial region; (7) AR9026 MDI data
covers the time of an X2.3 class solar flare which has been reported to cause
a sunquake; (8) A series of active regions and close quiet Sun regions analysed
identically are available for comparative purposes.

Figure 3 shows a quick look intensity image of an example data set with
three sunspots in AR8402 and AR8403. A total of 8 hours of high-resolution
MDI Dopplergrams are available for local helioseismology. For each set of active
region observations a link is provided to the Solar Monitor web page with quick
look maps in various observing spectral lines. External links to active region
and synoptic maps provide a good overview of the solar activity at the time.

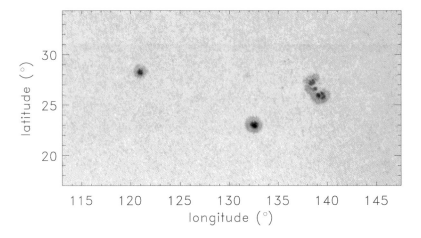

Figure 3: MDI continuum image from a particular HELAS data set showing three sunspots in active regions AR8402 and AR8403. A total of 8 hours of high-resolution MDI Dopplergrams are available for helioseismic analysis.

Important information has been collated for each data set: the Carrington latitude and longitude of the observation, links to sample headers, FITS header keyword definitions, and general information on the physical active region characteristics. These characteristics include the co-ordinates of the sunspot, the magnetic type, classification, the area, the longitudinal extent of the sunspot, the average umbral and penumbral boundaries and the number of sunspots present on the solar disk at the time of the observation. In some cases, far-side images of the active region are shown which give an indication of the active region's life span. This is important for a full analysis of the observations.

Data reduction

Most of the observations have been pre-processed and it is important to know exactly how. The webpage http://www.mps.mpg.de/projects/seismo/NA4/ DATA/data_red.html links to published papers about the instruments, the data processing pipelines and the known characteristics of the data. Papers describing the merging of GONG data and the onsite data reduction are available for direct download. A schematic of the pipeline involved in delivering GONG data products is linked. The paper by Scherrer et al. (1995) contains all the relevant instrument details for the MDI observations. The MDI website contains many more details in regard to image quality, duty cycle and other performance characteristics. Similar information for the other instruments is primarily available in the form of papers.

Most of the observations available have been further processed. In local helioseismology, tracking of the local solar surface in time is necessary to remove solar rotation. A particular latitudinal rotation profile is used to do the tracking and we provide each profile for each data set. In the majority of cases for the data provided by the website this is done using an azimuthal-equidistant projection and mathematical descriptions of the many possible projections that could be used are linked.

Data analysis tools

Currently, three data analysis software tools are available from the HELAS local helioseismology activity website at http://www.mps.mpg.de/projects/seismo/NA4/SW/: the ring diagram analysis pipeline, code to measure travel times and holography codes. Documentation for the codes is provided in the form of linked reports, papers and presentations. The actual computer codes are available for direct download. Here we describe the data analysis software tools available.

Data analysis tool 1: Ring diagram pipeline

Ring diagram helioseismology (Hill 1988) is based on the analysis of the local acoustic power spectra in wavenumber and frequency space. In planes of constant frequency the power is concentrated in rings. Analysing the changes in the shape of these rings provides information about anomalies in the sub-surface structure of the Sun, such as flows and sound speed perturbations.

The GONG ring diagram software package comprises all the binaries, scripts and support files needed to obtain the subsurface averaged velocity flows associated with individual sections over the solar surface using the ring diagram technique. It has been optimised to take GONG Dopplergrams as input, although it could easily be adapted to accept other helioseismic observations. Various authors wrote the individual parts of the pipeline, however the package was put together for HELAS by I. González-Hernández (National Solar Observatory) and has been made available to download from the HELAS local helioseismology web page. Extensive documentation related to installing and running the pipeline is also linked, courtesy of A. Zaatri (Kiepenheuer-Institut für Sonnenphysik) and T. Corbard (Observatoire de la Côte d'Azur), as well as presentations and relevant scientific papers.

Data analysis tool 2: Helioseismic holography codes

Helioseismic holography (Lindsey & Braun 2000) takes the acoustic amplitudes observed at the surface and computationally regresses them into the interior either forward in time (ingression) or backward in time (egression). The results give us an acoustic image of the solar interior.

The basic codes were written by C. Lindsey (Colorado Research Associates) to calculate the ingression and egression of pre-processed MDI Dopplergrams for helioseismic holography and are available to download. The HELAS holography web page gives the specific information required by these programs so that users can tailor their input data accordingly. The website also provides documentation about the technique by providing links to the scientific papers on the basic principles of helioseismic holography.

Data analysis tool 3: Fitting travel times

The goal of time-distance helioseismology is to calculate the travel time of wave packets observed on the surface over some distance. This is done by calculating the cross-covariance function between the signal at two points. The travel times are measured by fitting wavelets to the cross-covariance.

The codes that are made available on the HELAS local helioseismology website consist of IDL routines for measuring travel times from the cross covariance. Two methods are implemented: a Gabor-wavelet fit and a one-parameter fit. Example cross-covariances and instructions for running the code are included in the package. A full description of the fitting methods is given in Roth et al. (2007).

Modelling tools

Models and simulations are important for understanding and interpreting the results of helioseismic analysis. We have collated a selection of model and simulation source codes relevant for local helioseismology. Here we describe the modelling tools that are available from http://www.mps.mpg.de/projects/ seismo/NA4/MODEL/.

Modelling tools 1: Numerical simulation of wave propagation

The Semi-spectral Linear MHD (SLiM) code numerically simulates the propagation of linear waves through an arbitrary three-dimensional atmosphere (Cameron et al. 2007). From these simulations the interaction of acoustic waves with various inhomogeneities in the solar atmosphere can be studied.

The code is written in Fortran90 and documented by published papers. Instructions to run the code comes in a READ_ME file in the download package. Fortran libraries required to run the code are also provided. See the screen shot in Figure 4 for more details.

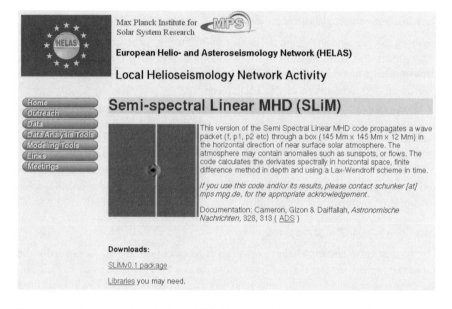

Figure 4: Screen grab of the SLiM higher level application tool webpage at http://www.mps.mpg.de/projects/seismo/NA4/MODEL/SLiM.html.

Modelling tools 2: Sunspot & fluxtube models

The interaction of acoustic waves with magnetic inhomogeneities of the solar atmosphere is an important subject of research. Codes which model various magnetic structures, sunspots and flux tubes, are made available from the website.

An IDL code creates a typical sunspot structure in a solar-like atmosphere (Khomenko & Collados 2006). The variable parameters are the magnetic field strength, the radius of the sunspot and the Wilson depression. A READ_ME file describes how to run the code and further information can be obtained by reading the listed publications.

A Fortran code that calculates magnetohydrostatic flux tubes according to the method of Steiner et al. (1986) is also provided. One input file contains all of the variable parameters which is easily modified. Extensive documentation to run the code is provided in the READ_ME file.

Modelling tools 3: Born travel-time sensitivity kernels

Travel-time sensitivity kernels are necessary to interpret helioseismic inversions. They describe the region of the Sun that a particular wave packet is sensitive to in the presence of some perturbation. In this case, the code to calculate the kernels is not supplied directly as there are numerous input parameters. Instead, Y. Saidi has developed a web interface and pipeline to expedite the calculation process. The computing resources are located at the MPS.

The interface allows users to specify input parameters for the kernel calculation (see Figure 5). The linear sensitivity kernels are based on a single-scattering Born approximation. Users have the choice between three different types of calculations as seen in Figure 5. The first set of calculations provides three-dimensional p-mode kernels for sound-speed perturbations (Birch et al. 2004). The second set of calculations provides three- dimensional travel-time sensitivity kernels for flows (Birch & Gizon 2007). The third set provides two-dimensional f-mode kernels for flows (Jackiewicz et al. 2007). Additional references are listed on the website for each calculation.

The results of the submitted calculation are available to download from a web address that is e-mailed to the user. An email is sent at each step of the computation. The first one will inform the user that the request has been received, the second one will notify the user that the computation has started and the third email will notify the user that the computation has ended, and if it is successful it will also carry the web address to download the results of the computation. Yacine Saidi developed this web interface and relevant details can be found in his Master's thesis (Saidi 2006).

Modelling tools 4: Ray tracing code

Time-distance helioseismology measures and interprets the travel times of wave packets propagating between two points located on the solar surface. The travel times are then inverted to infer sub-surface properties that are encoded in the measurements. Thus, it is useful to have some knowledge of the path the wave packet takes and the region of the Sun it passes through.

Codes are provided that calculate helioseismic ray paths written by A.C. Birch (Colorado Research Associates). There are three main codes which calculate the group time as a function of distance, the ray paths, the phase and group times along the path (Birch 2002). Documentation explaining the execution and function of the codes is provided.

Meetings and workshops

There have been two HELAS workshops and one splinter meeting dedicated to the HELAS local helioseismology network activity, all of which had a high international participation. The first workshop, *Roadmap for European Local Helioseismology* (http://www.oca.eu/HELAS/), was held at the Observatoire de la Côte d'Azur, Nice, France on 25-27 September 2006. There were 38 participants who gave talks on the status of local helioseismology. In light of this

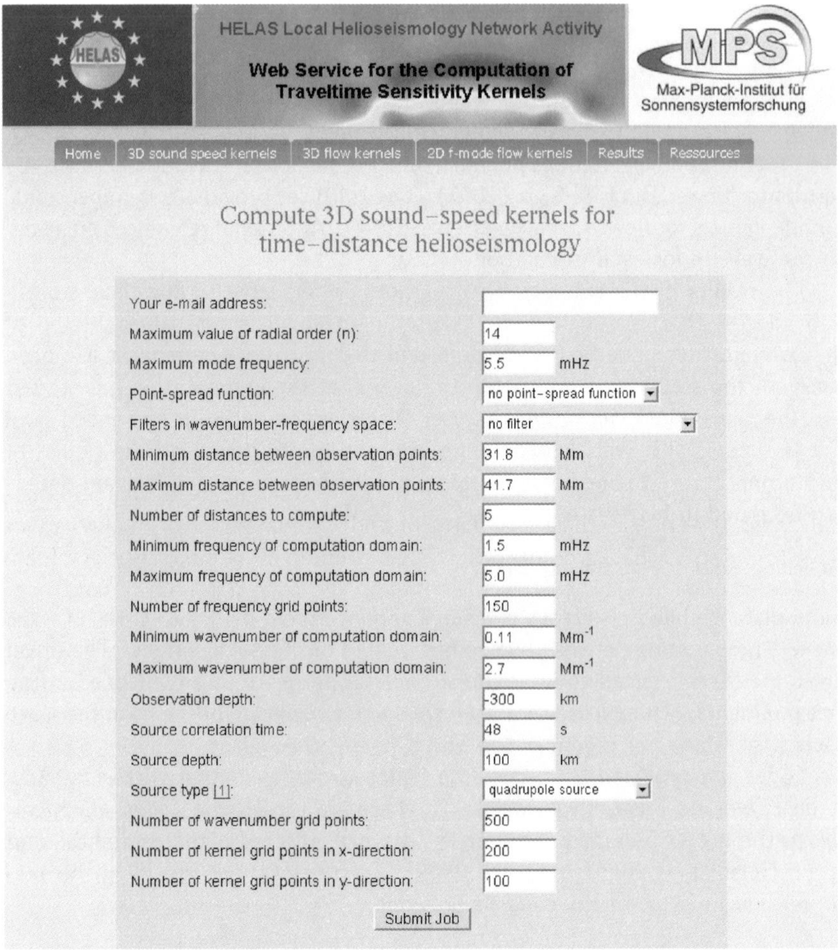

Figure 5: Screen grab of the web interface showing some of the input parameters to calculate 3D sound speed kernels for the computation of travel-
time sensitivity kernels. The interface is linked from http://www.mps.mpg.de/
projects/seismo/NA4/MODEL/tt_interface.html.

there were discussions about the direction local helioseismology should take, particularly in Europe. The proceedings consist of 28 papers and were published in the journal *Astronomische Nachrichten* Volume 328, Issue 3-4 accessible from the web site http://www3.interscience.wiley.com/journal/114173529/issue.

The second workshop was held at the Kiepenheuer-Institut für Sonnenphysik in Freiburg, Germany on 7-11 January 2008 (http://www.mps.mpg.de/projects/seismo/HLHW2/). Twenty-eight people attended by invitation and were divided into three teams, that covered the analysis of active region AR9787, numerical MHD simulations, and preparations for the Solar Dynamics Observatory. During the workshop Team 1 actively analysed the sunspot in AR9787 using MDI/SOHO observations made available on the HELAS local helioseismology website. Three main methods of helioseismology – time-distance, ring-diagram and helioseismic holography – were involved. This made it possible to directly compare the results for identical observations by eliminating any possible differences in initial data reduction. This had not been done before for a sunspot. Team 2 discussed issues related to the numerical treatment of wave propagation in the near surface layers of the Sun. Particular emphasis was placed on two topics: the need for a set of standard tests which numerical codes should reproduce as a form of code validation; and on the various ways to construct background solar models containing inhomogeneities. Team 3 made significant progress towards handling the large amounts of data from the Solar Dynamics Observatory (SDO) which will be launched in 2009. The team focused on implementing and discussing the Data Record Management System (DRMS), which is the software that will be responsible for managing SDO data. The DRMS is being developed and distributed by the Joint Science Operations Centre (JSOC) at Stanford University in the USA, and is also operational at the German Data Centre (GDC) for SDO. The analysis of AR9787 lead to interesting discussions which were continued at the International Space Science Institute meeting, *The origin and dynamics of solar magnetism* that was held in Bern, Switzerland on 21-25 January 2008. This has resulted in a forthcoming publication in Space Science Reviews (Gizon et al. 2008).

A splinter session devoted to discussing the HELAS local helioseismology deliverables was a part of the HELAS II International Conference - *Helioseismology, Asteroseismology and MHD Connections* (http://www.mps.mpg.de/meetings/seismo/helas2/) held in Göttingen, Germany on 20-24 August 2007. The suggestions and comments from the participants contributed to the selection of observations, software tools and models are made available on the HELAS local helioseismology website. The proceedings of this meeting are published in the *Journal of Physics: Conference Series* volume 118. In collaboration with the SOHO 19/ GONG 2007 meeting held in Melbourne, Australia on 9-13 July 2007, a Topical Issue of *Solar Physics* was also published and is listed in the following section. This Topical Issue places a strong emphasis on local helioseismology.

Judging from the resulting publications, the workshops fostered strong collaborations between the working groups active in local helioseismology in Europe and elsewhere around the world.

Publications

A total of 163 papers have resulted from the HELAS local helioseismology network activity. These papers appear in the following publications:

- *Astronomische Nachrichten*, Volume 328, Issue 3-4. Proceedings from the First HELAS Workshop, *Roadmap for Local Helioseismology* held at the Observatoire de la Côte d'Azur, Nice, France on 25-27 September 2006. The volume consists of 28 papers.

- *Solar Physics*, 'Helioseismology, Asteroseismology and MHD Connections', volume 251, 2008. This is a joint effort between the HELAS II International Conference and the SOHO 19/ GONG 2007 meeting. Many of the papers included here represent work presented at one or other of the meetings, but this Topical Issue was opened for general submission on their core topics. From the 43 papers in the volume, 26 are focused on local helioseismology.

- *Journal of Physics: Conference Series, Proceedings of the second HELAS international conference: Helioseismology, Asteroseismology and MHD Connections*, volume 118, 2008. The volume contains 11 local helioseismology papers out of 91 papers in total.

An additional paper has been accepted for publication:

- L. Gizon, H. Schunker, C.S. Baldner, S. Basu, A.C. Birch, R.S. Bogart, D.C. Braun, R. Cameron, T.L. Duvall, Jr, S.M. Hanasoge, J. Jackiewicz, M. Roth, T. Stahn, M.J. Thompson, S. Zharkov, 'Helioseismology of sunspots: A case study of NOAA region 9787', *Space Science Reviews*, submitted, 2008. This is a joint publication resulting from the second HELAS local helioseismology workshop.

Future activity

One more workshop is planned to be held in the first half of 2009. The observations, software analysis tools and modelling code will continue to be modified and updated to keep up with progressions in the field.

Acknowledgements

Many thanks to all contributors to the HELAS local helioseismology network activity including Charles Baldner, Aaron Birch, Rick Bogart, John Bolding, Doug Braun, Robert Cameron, Thierry Corbard, Tom Duvall, Wolfgang Finsterle, Shravan Hanasoge, Irene González-Hernández, Frank Hill, Jason Jackiewicz, Elena Khomenko, A.G. Kosovichev, John Leibacher, Charles Lindsey, Markus Roth, Yacine Saidi, Oskar Steiner, Mike Thompson, Thomas Wiegelmann, Amel Zaatri and Sergei Zharkov. Special thanks to all others from the HELAS local helioseismology nodes at the Observatoire de la Côte d'Azur, Max-Planck-Institut für Sonnensystemforschung, Kiepenheuer-Institut für Sonnenphysik and the University of Sheffield for their support.

References

Birch, A. C. 2002, PhD thesis

Birch, A. C., Kosovichev, A. G., & Duvall, T. L. 2004, ApJ, 608, 580

Birch, A. C., & Gizon, L. 2007, AN, 328, 228

Cameron, R., Gizon, L., & Daiffallah, K. 2007, AN, 328, 313

Duvall, T. L., Jefferies, S. M., Harvey, J. W., & Pomerantz, M. A. 1993, Natur, 362, 430

Gizon, L., Schunker, H., Baldner, C. S., et al. 2008, SSRv, accepted

Hill, F. 1988, ApJ, 333, 996

Jackiewicz, J., Gizon, L., Birch, A. C., & Duvall, T. L. 2007, ApJ, 671, 1051

Khomenko, E., & Collados, M. 2006, ApJ, 653, 739

Lindsey, C., & Braun, D. C. 2000, SoPh, 192, 261

Roth, M., Gizon, L., & Beck, J. G. 2007, AN, 328, 215

Saidi, M. Y. 2006, M.Sc. Dissertation, Universités Paris Sud XI, Paris

Scherrer, P. H., Bogart, R. S., Bush, R. I., et al. 1995, SoPh, 162, 129

Schunker, H., Gizon, L., & Roth, M. 2008, Proceedings of the Second HELAS International Conference, JPhCS, 118, doi:10.1088/1742-6596/118/1/012087

Steiner, O., Pneuman, G. W., & Stenflo, J. O. 1986, A&A, 170, 126